Dolls in Denmark —Then and Now

Dedicated to
Gunhild Nielsen,
my mentor and dearest Danish friend.

Inquiries should be addressed to
Oralu Corporation, 1009 Sitka Ct., Loveland, CO 80538.

ISBN 0-9631475-0-1

Library of Congress Catalog Card Number 92-90719

Luckey, Pauline
Dolls in Denmark – Then and Now

Printed and Bound in the United States of America
by Johnson Printing, Loveland, CO 80538

frontispiece

Doll of Her Majesty Queen Margrethe II,
commissioned for the Queen's fiftieth birthday April 16, 1990
by Kirsten Gynther Andersen.

Dolls in Denmark —Then and Now

by
Pauline Luckey

Oralu Corporation
Loveland, Colorado
1992

Forword and Acknowledgments

The story of dolls involves their related toys and accessories, and is a reflection of their place in history.

It has been impossible to entirely separate the presentation of the dolls from historical figures and figurines. Describing only the dolls would be telling only part of the story. The puppets, the jumping jacks, the coin banks and squeeze toys, the mascot dolls, and the Dam trolls (not dolls per se, but now accepted as collectibles within dolldom) are related, and are a part of the overall production of dolls in Denmark.

It is also quite impossible to present one factory's production of dolls and ignore its preceding or simultaneous developments. The doll industry of J.K. Kofod in Køge* was built upon earlier manufacturers' efforts, and the rubber balls, bath toys, and banks and animals were a part of the firm's enterprises. Similarly the the banks by DanCoin, and by Thomas Dam in Gjøl are parallel products of their dolls and trolls.

Doll houses and doll furniture are mentioned in the text only incidentally, as found in certain museum collections and in factory production, not as accessories being documented here. The study of either subject would become another vast research challenge, fascinating but not within the time and energy of this author and photographer.

Acknowledgments must be made to many wonderfully helpful and interested individuals, museum directors, private collectors, designers and manufacturers, doll artists and artisans who have generously shared their knowledge, their photographs and their dolls, their expertise and translations. Most of these acknowledgments appear within the text and with the illustrations. To every one of them the author and the photographer are greatly indebted.

The information in Chapter 14 is presented with acknowledgments cited here to many individuals. A special thank you is expressed to Mr. Bjarne Kildegaard, Ph.D., Curator, Dept. of Modern Culture, *Nationalmuseet*, NYF. RF. TID, DK-2800 Lyngby, for permission to use illustrations and quotations from the book *Dukken Margrete*, and to Mrs. E. Grolstadt who arranged our photographic session of the doll in the museum. More special thanks are due Mrs. Gunhild Nielsen who translated the text, and the author, Minna Kragelund, who attended conferences at the Nielsen home. The following doll authorities in Germany, R. and K. Knedel of Frankfurt am Main and Mr. G. Gauder of Stuttgart, are thanked for their assistance in identifying the doll from transparenies. This extended history of dolls in Denmark developed from the translation and study of the doll Margrete, as suggested by Mr. James Lee Young, former Executive Director, National Writers Club, Aurora, Colorado.

The studied materials could not have been faithfully presented without the work of friends in Denmark who have helpfully provided translations into English. Sincere thanks are expressed to Mrs. Kirsten Jexlev of Copenhagen, Mrs. Pamela Nedergaard of Odense, and Herluf and Gunhild Nielsen of Charlottenlund. Many delightful days were spent in that endeavor in the Nielsens' plant filled solarium, with renewals of tea breaks. The Nielsens have also supplied many needed facts, clues, and useful books which they translated, and family dolls which are documented in this book. Mr. Andrew Staby of Loveland, Colorado provided many needed instant translations.

Research opportunities were enriched at the Euro Toy Show and Sale at the Hotel Vis-a-Vis in Billund in October 1989. This twice yearly international event is arranged by Kurt Strebel, affiliated with Nellemann and Thomsen *Kunstauktioner*. His suggestions and introductions to exhibitors were appreciated assistance to us there. The 1990 auction catalogues he has sent contain examples otherwise unknown.

Acknowledgments are made to those who carefully read and critiqued parts or all of the text. Those include Coloradans Mildred Camp, Shirlee Glass, Martha and Andrew Staby, and especially my best critic, my photographer husband, Don Luckey. In Denmark, Camme Bache in Praestø, Niels Dam in Gjøl, Mrs. Else Madsen at DanDolls, and Mrs. Kirsten Martin Stadelhofer at Legoland A/S have carefully critiqued those chapters resulting from graciously given interviews. Mrs. Gunhild Nielsen gave much excellent council after studying chapters of the manuscript.

To Eileen van Baren, a Loveland artist, special thanks are expressed for her sketches which begin each chapter, and for many others which appear throughout the text. And to Fumiko Ino, Tokyo, a thank you for the sketch of the author appearing on the back cover.

Parts of Ch.8, text and illustrations, have been published in *International Doll World*, Oct.1991, and reappear in the book, courtesy of Rebeka Montgomery, Editor.

Grateful appreciation is due to my family, to our daughters and their spouses for moral and financial support, and to Don Luckey for tireless doll hunts, his photography, his critiques, and especially for his encouraging t.l.c. in support of the project through many years.

*The reader will encounter the diphthong "o" in names of towns such as Køge and Gjøl, and may consider the Danish sound to be the equivalent of the German umlauted "o", or the English sound of the combination "oe".

Table of Contents

Introduction

Doll history is forever fascinating, and it will never be complete. "All of doll history is transition; transition from one era to the next; transition from one medium to the next" [1]. Books by numerous authors in the United States, Germany, France and Holland give almost nothing about dolls in Denmark. Two books by Danish authors, *Gamle Dukker* (*Antique Dolls*) by Susanne Tvermoes [2], and *Det Legede Vi Med* (*That's What We Were Playing With*) by Jens Sigsgaard and Ib Varnild [3], and the English author Madeline O. Merrill's *The Art of Dolls* [4] are very helpful.

This author and photographer team hopes to introduce the rarity and beauty of Danish dolls, antique to contemporary, to readers and collectors. Our hunt for dolls and doll history has been pursued on numerous trips to Denmark since 1961. This volume will present our research there and elsewhere on dolls in Denmark, then and now.

Danish doll research invariably includes related toys, teddy bears and other stuffed animals, jumping jacks called *Spraellaemaend*, doll houses, carriages and beds, and coin banks, figurines, candle sticks and other items. Those will be described as they occurred in museums, antique shops and flea markets, and in manufacturing. They are not an integral part of this study, but are included as encountered in the hunt.

We found that more Danish dolls have been made than anyone, even the Danes themselves, had imagined.

Danish friends, collectors, artists, and museum and doll factory personnel have provided valuable information and sometimes mere clues which have revealed wonderful mysteries.Their names will be included in discussions and acknowledgments. Museums where dolls may be seen in the three areas of Denmark are discussed in Chapter 13.

This story of Danish dolls embraces Denmark's history, traditions, folklore, and geography. Historically the very early kings and queens are represented in the beautiful contemporary creations of Kirsten Gynther Eriksen. Traditional folk costumes by several firms and by doll artists will be shown in a great variety. Scandinavian folklore includes the Dam Trolls of Gjøl and lovable Danish *Nisser* (Christmas elves), antique and modern.

Denmark is a country of three major areas. Jutland is the peninsula extending north of West Germany, and the two large islands are Funin and Zealand. The autonomous islands, the Faroes and Greenland, and the many very small islands in Danish waters total 500. A map of Denmark (Ill.1) shows the location of doll firms, doll artists, and museums which enrich this story. Each of more than thirty places will be encountered in the text, showing the extent of Danish doll activity.

The earliest doll whose story has been documented, in 1967, is "*En dukke fra 1830 - 40*" ("*A doll of 1830 - 40*") [5]. An oral translation of that chapter was graciously provided by Mrs. Kirsten Jexlav, of the library staff of the Danish National Museum in Copenhagen, in October 1989. It is a study of a doll which belonged to a girl in Copenhagen in the 1830s. Although it has been called a "Fanø doll", its origin had nothing to do with the Island of Fanø. It appears to have been made in Sonneberg, Thuringia, historically one of the doll centers of Germany. The author stated that although one cannot be one hundred percent certain, the doll could have been exported from Sonneberg, where dolls with papier mâché heads were made as early as 1806. This doll's papier mâché head was painted like those of early German

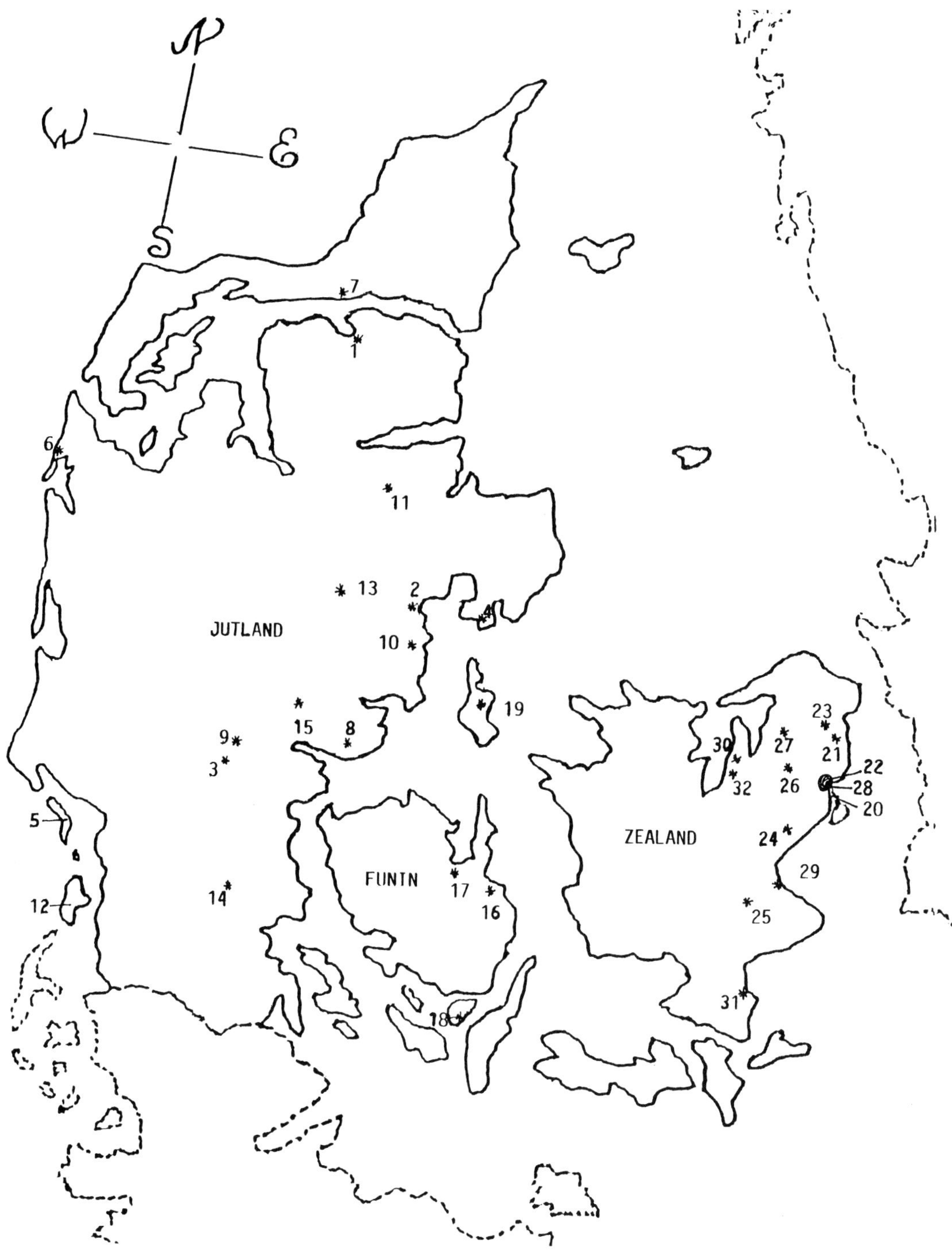

(Ill.1)

Where Are The Dolls In Denmark?

CITIES AND TOWNS ON, AND ISLANDS NEAR

JUTLAND:

1. Aalborg
2. Aarhus
3. Billund
4. Ebeltoft
5. Fanø
6. Fjaltring
7. Gjøl
8. Hornsyld
9. Legoland
10. Odder
11. Randers
12. Rømø
13. Silkeborg
14. Toftlund
15. Torring

FUNIN:

16. Nyborg
17. Odense
18. Tasinge
19. Samsø

ZEALAND:

20. Amager
21. Charlottenlund
22. Copenhagen
23. Gentofte
24. Glostrop
25. Herlev
26. Holte
27. Horsholm
28. Kastrup
29. Køge
30. Ølstikke
31. Praestø
32. Valby

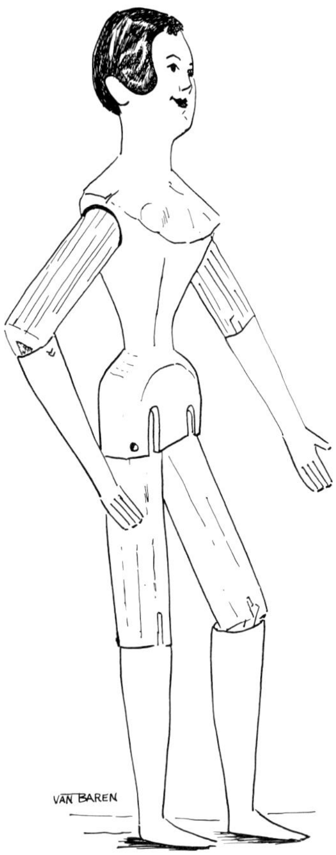

Ill. 2 a

Ill. 2 b

dolls. The molded hair was styled in the fashion of 1830, no later than 1840. Dolls with porcelain heads, which were made in Copenhagen in the 1840s, will be presented in Chapter 1. However, according to the 1967 study, no doll industry existed in Denmark in 1830.

This early doll is almost 10 in (26 cm) tall. The papier mâché head, and arms and legs of fir are on a body of elder wood, a kind of poplar. The articulation is visible in the sketch of the doll (Ill.2a).

The costume and the basket carried on the doll's arm suggested that it represented a peasant girl of 1830 (Ill.2b). The homemade clothing was sewn on. Underclothes of cotton and a white linen petticoat were worn beneath the two piece black wool dress. The front opening of the blouse was edged with red silk, and the long sleeves with white edging. Ribbon trimmed the skirt. A white apron was worn over the black dress, typical of several dolls' costumes of the following decade. A scarf of rose wool and silk blended fabric covered the doll's hair. Worn at the neckline was a triangular scarf made of several seamed together bits of expensive silk taffeta, still a colorful print of blue, green, black and white. This was wrapped to the back and was sewn onto the doll. Perhaps it was the wrapped head scarf and the neck scarf, and the dark wool dress which suggested the Fanø doll identity? Dolls in regional costumes of Fanø, presented in Chapter 3, will show similar details.

Following the presentation of doll museums in Denmark in Chapter 13, the star of the Danish National Museum, an English wax doll, "Dukken Margrete", portrays life in a wealthy Copenhagen family in the 1860s. That doll's clothing and accessories are in vivid contrast with this little papier mâché and wooden lass in humble peasant dress.

Between those two dolls' stories, several chapters will present Danish dolls according to materials from which they were made. The variety, papier mâché, wood, porcelain, composition, celluloid, polyvinyl and hard plastic, cloth, and sculpting materials, and their beauty are startling. Two chapters will discuss the fascinating histories and production of two factories in Jutland, the Dam Trolls of Gjøl, and the DanDolls of Hornsyld.

A summary of our research through May 1989 revealed deficiencies in our findings, prompting our return to Denmark in October 1989. At the "Euro Toy Show and Sale" at Legoland A/S we met more collectors and artists, and we photographed and documented more dolls. Before we returned we hosted a coffee hour at our hotel for the doll artists we had met in Denmark. They have willingly shared their stories and their dolls which will appear in the text, and each will be profiled in Chapter 12.

(2 a) The 1830 - 1840 doll showing articulation.

(2 b) The doll showing outer clothing. Sketch by Eileen van Baren.

This presentation is not complete. The story of Danish dolls, some virtually unknown but all remarkable and all fascinating, historically and artistically, needs to be shared!

We anticipate that our Danish dolls will be displayed in the Danish Immigrant Museum under construction in Elk Horn, Iowa, U.S.A. There, in the Changing Gallery, the Educational Collections will be organized according to exhibits of seasonal or "special themes...Christmas traditions, lace making, stamps, or dolls", wrote June Sampson, Director. Dolls in the Ora Elizabeth collection (of author and photographer) which were found in Denmark are seldom seen in this country. With other dolls given to the museum (described in Ch.3 and 6) they will be a significant addition to dollology.

(Ill. 3) Danish immigrant museum under construction in Elk Horn, Iowa, from architectural drawing. Courtesy of June Sampson, Director.

Chapter 1

Danish Porcelain Dolls – Then and Now

Royal Copenhagen Porcelain Dolls of the 1840s

A pair of exquisitely dressed, brown haired porcelain head dolls was displayed in the Royal Copenhagen Porcelain shop at Vesterbrogade 8, Copenhagen at Christmas time, 1978. The glazed porcelain heads (which are called china heads in the U.S.A.) were replicas made from the 1840s mold said to represent Queen Charlotte Amalie, the wife of Christian VIII (reign: 1839 - 1848). The doll is named "Queen Amalie"* or simply "Amalie". Three heads were available: 1.287, Amalie, 5 1/2 in (14 cm), 1.055, a girl with brown hair, 3 3/4 in (9.6 cm) and another 1.055, a boy of the same size, with lighter hair (Ill.4). A wonderful Christmas gift to ourselves was the Queen Amalie head (Ill.5a,b). The pattern and Danish directions for the body were gifts from our hosts.

This Amalie replica head is characterized by center parted dark brown hair combed over her ears and back into a twisted bun above her hairline, a serene closed mouth with a hint of a smile, brown eyebrows and painted blue eyes. The soft red eyeliner above the eyes instead of eyelashes, and the red dots in the corners of the eyes, tear duct dots, are typical of delicately painted china heads. Soft red dots mark the nostrils. A soft coral-red lightly rouges her cheeks and her chin. Most unforgettable is the very lovely, very natural flesh tone of the skin.

The company's marks were made inside the back of the shoulder plate, making identification of the old dolls difficult when compared with German porcelain heads of the same period.

Our Amalie head is marked with an incised "A"// "blue stamped wavy lines", brown "ggx"/, blue "DENMARK//11" (Ill.6). The middle line of marks is in line with the three sew holes on the shoulder plate. The three wavy lines representing Royal Copenhagen Porcelain, acccording to the firm, represent three of the bodies of water around Denmark: the Great Belt between Zealand and Funin, the Little Belt between Funin and Jutland, and the Sound. This is the latest of the blue marks used by this firm; earlier marks are shown with descriptions of the antique porcelain dolls.

* Only dolls' names cited in museums or private collections, not dolls' names given by firms, are in quotation marks, and those only the first time used within the text.

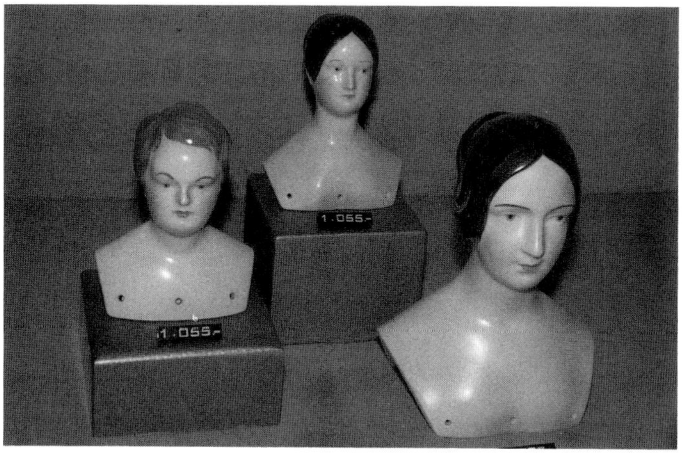

Ill. 4

Ill. 5 a

Ill. 5 b

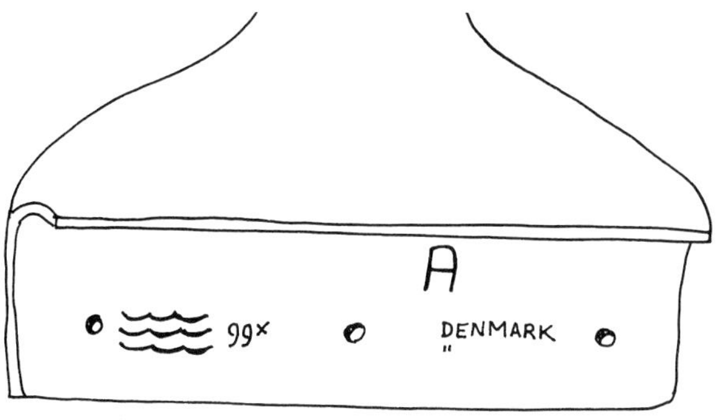

Ill. 6

Ill. 7

Ill. 8

Ill. 9

(4) Three replica heads by Royal Copenhagen Porcelain. Courtesy of Royal Crown Copenhagen Porcelain.

(5 a,b) Amalie head in author's collection.

(6) Marks inside shoulder plate of Amalie. Eileen van Baren sketch.

(7) Profile of doll made of papier mâché and of porcelain. E. v B. sketch after illustration, Fawcett, p.145.

(8) Amalie in profile. E. v B. sketch after photo, Dröscher, p. 46.

(9) From a painting by Ekersberg, "Study of a Nude in Front of a Mirror", showing the classic hair style.

An 1840s porcelain Queen Amalie head displayed in the Royal Danish Porcelain Shop in Copenhagen stimulated our need to know the history of the company which made those beautiful doll heads. The Royal Copenhagen Porcelain Co. was founded in 1775. From Queen Juliane Marie the firm acquired the patronage of the Danish Royal Court. Toward the middle of the next century the firm was making doll heads; twelve molds represented men, women and children. Probably much better known than the dolls are the Christmas plates, begun in 1908 and continuing in a limited annual pattern. Their popularity has prompted the addition of a mini-Christmas plate, 2 in (5.5 cm) in diameter, the 8 oz Christmas cup, and a tree ornament, all made in the pattern of the plate.

Although the annual Christmas porcelains are produced in limited numbers, the replicas of doll heads are not. Those three replicas were available at the Royal Copenhagen Porcelain shop in 1989. Decamp Classics, Inc. in Scotch Plains, New Jersey made the three heads available in December 1981. But the antique porcelain dolls are scarce indeed. Numerous doll museums in the U.S.A. were queried; only one has responded documenting its very rare Danish porcelain doll. The Margaret Woodbury Strong Museum at Rochester, New York has a doll head, a black haired boy with the early mark of the three blue vertical lines.

Photos of two old dolls are seen in doll encyclopedias by the Colemans. Both volumes show a lovely china head which has the "brown-black hair...intertwined into a bun placed at the center back of the head" [6]; the classic styling has also one wispy curl behind each ear. Three vertical lines with a dot below the middle line mark this doll head. A rarely seen man's shoulder head with molded black hair, with the mark of the doll just described, is shown in Volume II [7].

Photographs are found in some of our European doll books [8, 9, 10, 11, 12]. A porcelain head doll, 21 5/8 in (55 cm) tall, is shown in two photographs, a close-up profile and a full length front view, in Elke Dröscher's book [13]. She looks like the Amalie doll, but is dressed a bit less regally in a long skirt and fitted hip length blouse, a *Hauskleid* (house dress), of printed cretonne. Over the large lace collar she wears a locket on a ribbon.

We have found the most complete source of illustrations to be Merrill's book from London [14] and from Hobby House Press in Maryland. Three views [14 a,b] of a doll head like the lady doll described by Colemans are shown. In the colored photograph the brush marks in the hair are very visible. The early mark identifying this style contains the three blue vertical lines with a dot beneath. The same mark identifies another rare doll head on a rarer pegged wooden body [14 c]. The 16 1/2 in (41.9 cm) doll has blue eyes; her hair exposes the ears and is drawn into a coil in back. Most surprising to find is [14 d] the flange head on the so-called Motchmann doll body of 1857, with black brush marked pate. This 11 in (27.9 cm) doll is identified by a recognized mark of the Royal Copenhagen firm, just three blue vertical lines. Another doll head has brown hair in an unusual style, falling in long loops in front of the ears and then drawn beneath the exposed ears to the bun in back. "In rare instances, brown haired dolls similar to the one shown have been found with the marks of the Royal Danish Manufactory of Denmark" [14 e]. A similar papier mâché head on a doll of the 1820s, described by a pioneer doll researcher and writer, Clara H. Fawcett [15], "is also found in china....Queen Victoria wore her hair in this manner in 1840, although the style was not new at that time". Sketches of profiles of those dolls with hair styles worn by Queen Victoria and of Queen Amalie are included (Ill.7, 8).

On the cover and within the text of the book *Danish Painting and Sculpture* is shown a painting by C.W. Ekersberg (1783-1853). He is called "the Father of Danish Painting" [16]. The painting, ca 1832, "Study of a Nude in Front of a Mirror," is reproduced (Ill.9) with permission of *Den Hirschsprungske Samling* (the name of the collection). It shows the back and the mirrored front of the classic hair style, similar to the doll Amalie's coiffure, but with exposed ears.

On our earlier visits to the Legoland A/S Doll Collection at Billund, Jutland we were delighted to see old Danish dolls and doll heads. Returning in 1989 we were shown the entire museum by Kirsten Martin Stadelhofer who graciously allowed photographing the dolls. Amalie and another named "Signe" are that museum's first old Danish porcelain dolls documented here.

The doll Amalie (Ill.10 a, b) wears a long sleeved black blouse decorated with lavish embroidery and rose colored fringe along a deep V and a jabot. Her floor length skirt is grey, yellow and brown plaid. A full white apron is slightly shorter than the skirt. Amalie, dated 1844, is just over 22 in. (56 cm) tall. The doll Signe (Ill.11 a,b) is the adolescent, 14 1/8 in (36 cm). She is exquisitely dressed in a floor length gown of light rose-pink flowered fabric. White lace edges the wrist length sleeves, the center front bodice, a skirt pocket, and the neckline accented with a pale pink ribbon rosette. A sheer stole gracefully drapes her shoulders and wraps around her wrists.

Three more Royal Copenhagen dolls are included in a wonderful collection obtained by Legoland A/S in 1989 from Copenhagen. Although they were not yet ready for public display, we were honored with a private viewing by Kirsten Stadelhofer who has kindly provided photographs and documentation to be included here.

Readily recognized is the 21 5/8 in (55 cm) brown haired Amalie (Ill.12 a,b), approximately the size of the museum's first Amalie. Her arms and hands are porcelain; her body is leather. She wears a low U necked gown of dark green, a lace ruffled apron of brown printed fabric, and over her shoulders an embroidered and fringed stole. Her necklace appears to be a triple strand gold chain.

Ill. 10 a

Ill. 10 b

Ill. 12 b

Ill. 12 a

Ill. 11 a

Ill. 11 b

Ill. 13 b

Ill. 13 a

(10 a,b) Legoland A/S doll Amalie, full length and close-up.
(11 a,b) Legoland A/S doll Signe, full length and close-up.

(12 a,b) Amalie, new acquisition by Legoland A/S. Full length and close-up. © Courtesy of Kirsten Martin Stadelhofer.

(13 a,b) Tante Luise Moltke, at Legoland A/S. Full length and close-up. © Courtesy of Kirsten Martin Stadelhofer.

Ill. 14 a

Ill. 14 b

Ill. 15

The doll "Tante Luise Moltke" (Ill.13 a,b) is slightly taller than 20 in (50 cm). Her brown hair is worn in waves covering her ears and then twisted into a chignon high on her head. As unusual as the hair style are the arms and legs, skillfully carved of wood. Details of the eyes and eyebrows are delicate as those of Amalie, but her lips and cheeks are darker red. She wears her original costume, a jacket fitted to the waist and then flared, which buttons all the way up the front. The floor length skirt flares into a deep ruffle of narrow pleats. These two dolls are marked in blue: a dot at the left of the middle one of three wavy horizontal lines.

The last of this trio at Legoland A/S, "Alexi" for "Alexandrine" (Ill.14 a,b), is smaller, 10 5/8 in (27 cm). This *drengedukke* (boy doll) wears a tailored white cap and a long white dress belted below the yoke. Both are made of firm cotton fabric. Toes of the bare feet show below the dress. This doll is the only small sized porcelain doll which has been located and documented, and is a valuable addition to the Legoland A/S collection and to this presentation.

At what age was this long dress worn by a small boy abandoned for trousers in the 1800s? The answer was found in an 1830 classic book, *Flugen til Amerika* (*The Flight to America*) [17], in which the beginning stanzas translated as follows:

When I was only a little one
and had begun school,
and got boots on both legs,
and thrown away my *buksekjole* (long dress)
I was not as calm as I am now.....

The porcelain doll heads of the Berlin and Dresden firms in Germany were made during the same period. They have similar beautiful flesh tones but may have dark blonde, black or brown hair.

A Berlin made *Konigliche Porzellan Manufaktur* (K.P.M.) doll head "Lorentze" (Ill.15) was photographed at Legoland A/S. It is included for comparison with the Danish made doll heads. The similarities of the German and the Danish porcelain heads are apparent in the hair and silhouette. The coiffures of a mother and an older girl are shown in illustrations from *Die kleine Anna* and from *Die Puppe Anna, Ein Buch für Mädchen*, Berlin, ca 1840 in Elke Dröscher's book [18]. It is readily seen that similar hair styles were worn in both countries in that decade.

Another place to see lovely old Danish porcelain head dolls is the *Dukke Museum*, Kronprinsessegade 6, Copenhagen, the private collection of Mme. Margot Siesbye. A Queen Amalie doll is displayed with other rare and beautiful dolls in a courtyard scene (Ill.16). She wears a floor length gown of pale yellow silk and an elegant overskirt of embroidered sheer white fabric, topped by a waist length cape.

Ill. 16

(14 a, b) Alexi, Legoland A/S boy doll boy doll. Full length and close-up. © Courtesy of Kirsten Martin Stadelhofer.
(15) German K.P.M. doll head at Legoland A/S.
(16) Queen Amalie, photo courtesy of Mme. Margo Siesbye.

Mme. Siesbye's latest acquisition in May 1989 was another Royal Copenhagen doll (Ill.17). She was pleased to let us photograph and document it. This doll is brown haired in the fashion worn by the young girls, and has a lighter skin tone but darker features than the others described. Her mouth and her cheeks are much darker; both black and red eyeliner highlight her deep blue eyes. Her plaid silk dress is shorter than those worn by the others, and exposes her high topped shoes. Very full sleeves are gathered below the elbows, and the collar is heavy lace. A gold button at the front neckline repeats the gold of her beads and bracelet. The youthful silhouette of this girl doll is emphasized by a flowing silk sash tied in a big bow.

Our own Queen Amalie doll head is on a cloth body with leather arms, hands and shoes made by a Missouri friend. The doll's floor length gown of light blue antique satin was simply styled (Ill.18). A lace trimmed overskirt of old cotton batiste and a cape of royal blue velvet were suggested by the costume of Mme. Siesbye's Amalie. Undergarments of old white cotton and lace cut from the Danish firm's pattern completed this doll's costume sewn by the author.

Fortunate are the doll collectors who can view the Royal Copenhagen porcelain dolls and doll heads in the Rochester, New York and the Legoland A/S and Copenhagen museums. Most frequently displayed are the Queen Charlotte Amalie dolls. Two girl dolls, a boy doll, and Tante Luise are variations to be seen in Denmark. Although dark brown is the most usual hair color on the lady dolls, the *drengedukke* and the boy doll head have black hair, and the replica boy head is made with lighter hair. The antique porcelain heads made by the Berlin and Dresden firms have light or dark brown or black hair, and the same beautiful flesh tones. The placement of the firms' marks inside the back shoulder plate makes it difficult to distinguish the German from the Danish dolls.

Ill. 17 b

Ill. 17 a

Ill. 18

(17 a, b) Mme. Siesbye's girl doll. Full length and close-up.
(18) Amalie, replica doll, dressed by author.

Bing and Grondahl's "Doll of the Year" 1983 - 1989

Next door to the Royal Copenhagen Porcelain shop in Copenhagen is the Bing and Grondahl shop, another treasure trove. The brothers Meyer Herman and Jacob Herman Bing, and the artisan Frederik Vilhelm Grondahl had founded this second firm in 1853. Both firms continue to produce porcelains of the finest quality. Both produce the annual Christmas porcelain plates; the Bing and Grondahl series was begun in 1895, thirteen years before the Royal Copenhagen series was begun.

The firm of Bing and Grondahl has created a unique series of collectible contemporary dolls called "*Arets*

Ill. 19

Ill. 20

(19) Mary, 1983, first of series. © Courtesy of Copenhagen Collectibles.

(20) Else (12 1/4 in), 1984, with Else figurine (8 in). Courtesy of A. Andersen, Antik Sofie-Amalie.

dukke" ("The Doll of the Year") (Ill.19). These were introduced in 1983 and were made only through 1989. Each is stamped with the year of the issue and has been limited to that year's production. All were inspired by popular figurines which the firm had made from 1900.

These dolls have fine glazed porcelain heads with molded hair, arms, and legs. The bodies are firmly stuffed cotton fabric. A "lifelike complexion has been achieved with the addition of not less than 24-Karat Gold". [19].

The first annual doll, Mary, 13 3/4 in (35 cm), is a sweet solemn child holding her own small cloth doll (Ill.19). She was introduced in the spring of 1983. Charming in a long rose printed white dress, she wears a little rose trimmed cap over her center parted blonde hair. The figurine of Mary, one of the most popular, has been made since 1900, and was available at the Bing and Grondahl shop in October 1989.

Else is the second doll of this series, 1984 (Ill.20). This demure little girl doll, 12 1/4 in (31.2cm), has dark blonde hair styled with bangs over which she, too, wears a little tight cap. Blue and white embroidery decorate the blue dress and cap, and eyelet embroidery trim on her petticoat shows beneath the long skirt. A tan shoulder bag personalized with her name completes her ensemble. Else and Mary were each issued for $250.00; each doll came with her own wooden stand.

Trine, the doll of 1985, is a 12 in (30.5cm) preteen girl titled "Youthful Boldness". Her decollete gown of flower patterned white is fashioned above her ankles, revealing lace edged pantaloons. This golden haired miss wears saucy curls and an expression of anticipation — for whom?

For the 1986 boy doll Hans, of course (Ill.21). Is not the shy lad offering Trine a bouquet of roses? The only boy doll in the series, 13 3/4 in (35 cm), he is the first brownette; his eyes also are brown. He wears a navy blue suit with long trousers, a short, double breasted jacket with crisp white collar and cuffs, and a plaid bow tie.

And now is the 1987 doll, Marianne, also interested in meeting Hans? This blonde lass, 13 in (33 cm), wears a shorter dress of blue printed white trimmed with blue ribbon bands at the cuffs, waist, and near the hem. In her left hand she carries a basket of flowers, in her right hand a few blossoms (Ill.22).

Sophie, the 1988 doll, 13 in (33 cm), is dressed as a cook, wearing a tall white chef's hat on her short blonde hair and a lace trimmed white apron over her striped cotton dress. She holds a covered tureen of the classic blue-on-white Danish design (Ill.23).

Josephine, "On the Threshold to Adulthood", is the last doll of the series, 1989. Her lace trimmed ankle length gown is ruffled at the neck, sleeves, and hem. A wide pink ribbon sashes her waist; a white book is held in her left hand. This doll is 14 1/2 in (37 cm) tall. Her brown hair softly rolled over her ears emphasizes her brown eyes (Ill.24).

This series of dolls beautifully portrays nine Danish children, from the wee Mary tenderly holding her own doll to the grown up Josephine. At the Bing and Grondahl shop a few of the last three dolls were displayed in 1989. And occasionally some of the dolls or figurines appear in antique shops or at toy fairs; we bought the figurine Mary in an antique shop. It would be difficult to choose a favorite. All are beauties! A serious collector would like all of them and also the figurines which had inspired these charming dolls made in the 1980s.

Ill. 21

Ill. 23

Ill. 22

Ill. 24

(21) Trine, 1985, and Hans, 1986. Courtesy of Alice Thorgaard.
(22) Marianne, 1987. Courtesy of Copenhagen Collectibles.

(23) Sophie, 1988 doll. © Courtesy of Copenhagen Collectibles.
(24) Josephine, 1989. Last of the series. © Courtesy of Copenhagen Collectibles.

Chapter 2

Kirsten Gynther Eriksen's Historical Dolls

The serenity of the regal Queen Charlotte Amalie, wife of King Christian VIII (reign: 1839 - 1848) was beautifully interpreted by the Royal Copenhagen Porcelain firm in 1840.

Equally regal kings, queens, and naval heroes of an earlier age are the bisque porcelain creations of Kirsten Gynther Eriksen made 145 to 150 years later.

This contemporary artist, a former teacher of gymnastics, lives near Nyborg, Funin on a large farm with orchards of 12,000 cherry trees. She has thoroughly researched the history of Denmark in museums, libraries and castles. "When I found that there were no dolls representing Danish historical persons...I began making Hans Christian Andersen and Christian II. Later I made many others." [20].

"A distinguished society" [21] was how the press heralded a display of her dolls in June 1988 at the Valdemar Castle, built by King Christian IV, on Tasinge Island south of Funin. In that elegant presentation of royalty, King Christian II and his youthful queen, Elizabeth of Habsburg, (Ill.25) were the earliest pair (reign: 1513 - 1523). This child bride was described as the most beautiful of the group in her gold brocaded robe lined with dark green velvet and shining with gold decoration. Ornamentation of her head piece and neckline are shown in Ill.26.

"The nobility (during the Italian Renaissance, 1500 and on) wore rich silks, satins, taffetas, velvets and cloth of gold or silver. Most of the fabrics were in strong colours, and black velvet was used a great deal. As foreign travel and trade increased, so did the variety of fabrics and the means of dying them" [22]. These rich fashions were, of course, adapted by Scandinavian royalty.

King Frederick II (reign: 1559 1588), the son of Christian III, was called Denmark's "'Renaissance King' par excellence" [23]. He strongly maintained the sovereignity of the seas and was a supporter of the sciences. Sofie of Mecklinburg, his queen, wears the simplest costume of all the women dolls, a stately black velvet dress adorned with a heavy golden chain. White lace cuffs, the attractive lace cap, and the stiff white neck ruff emphasize that elegant simplicity. The ruff, readily recognized as Dutch, was worn also by French gentlemen in the late 16th century, by both men and

Ill. 25

Ill. 26

Ill. 27

Ill. 28

(25) King Christian II and Queen Elizabeth of Hapsburg.
© Courtesy of Kirsten Gynther Eriksen.

(26) Details of costume of Queen Elizabeth.

(27) King Frederick II and Queen Sofie of Mecklinberg.
© Courtesy of Kirsten Gynther Eriksen.

(28) Details of armor of King Frederick II.

Ill. 29

Ill. 30

(29) Kirsten Munk, King Christian IV, and their daughter, Leonora Christine. Courtesy of Kirsten Gynther Eriksen.
(30). Leonora Christine with 70 year old Christian IV.
© Courtesy of Kirsten Gynther Eriksen.

women in Elizabethan England, and in the Spanish court as late as the 17th century [24]. The pewter breast-plate of the doll, Frederick II, was molded by the artist and darkened with acid (Ill.27); authentic details are shown in Ill.28.

More figures represent periods in the life of Christian IV (reign: 1596 - 1648), set apart as "Denmark's most renowned King—the one all Danes know best", who inherited the two Kingdoms of Denmark and Norway [25]. His first wife Anna Catherina died young; his second marriage, when he was 40, lasted only ten years. This virile king is flanked by Kirsten Munk whom he considered the most important of the women in his life, and their daughter Leonora Christine (Ill.29). Flamboyance characterizes his dress in the large lace trimmed collar, more lace at the top of his boots, and the plumed hat. Kirsten Munk is resplendent in velvet with much gold braid and decoration on the stomacher which is the V shaped bodice, and undersleeves. Her high, stiff lace edged collar reflects a fashon of both the Elizabethan and the Spanish courts [26]. Bright red velvet is worn by their daughter, the narrow gold braid trim in contrast with wider trim on Kirsten Munk's costume. Leonora Christine's hat, too, is plumed.

In the second pair of these figures (Ill.30), Christian IV is the greying 70 year old king who had lost one eye in a naval battle with the Swedish forces. He still wears a lace collar and cuffs, and more lace atop his boots. Above his white eye patch he wears a small white lace cap, always the elegant king. An older but very beautiful red haired Leonora Christine wears a gown as richly trimmed as her mother's in the earlier group; again she wears a plumed hat.

Niels Juel (1629 - 1697), a famous admiral during Christian IV's time (Ill.31), is dashing in red: dark red trousers and brighter red coat enhanced by a rose-red sash. An ornate pierced cross hangs from a pale blue silk ribbon at his neck. Details in dress worn by his king, the lace trim on the plumed hat and more lace atop his boots, are elaborate touches on the admiral's costume.

The group of Christian IV dolls are second only in popularity to Hans Christian Andersen. This teller of tales is readily recognized by his top hat, flowing cape and walking stick. The artist is shown with this doll holding a Gothic lettered book of one of his stories, smaller than a match box, which was made exclusively for him (Ill.32). To accompany Andersen, Kirsten Eriksen has made a doll representing the "Princess and the Pea", on her bed with twenty mattresses (Ill.33). This little princess was created using a commercial mold of a three-faced doll head. In firing, the size of the mold is reduced slightly. Several successive moldings and firings were needed to obtain the wee doll head needed. The knob on top of the head is concealed inside her little gold colored crown; turning the knob reveals a sleeping face. Her long golden blonde hair and pink silk nightgown make this an endearing doll. Recently the artist made another figure of Hans Christian

(31) Admiral Niels Juel. © Courtesy of Kirsten Gynther Eriksen.

Ill. 31

Ill. 32

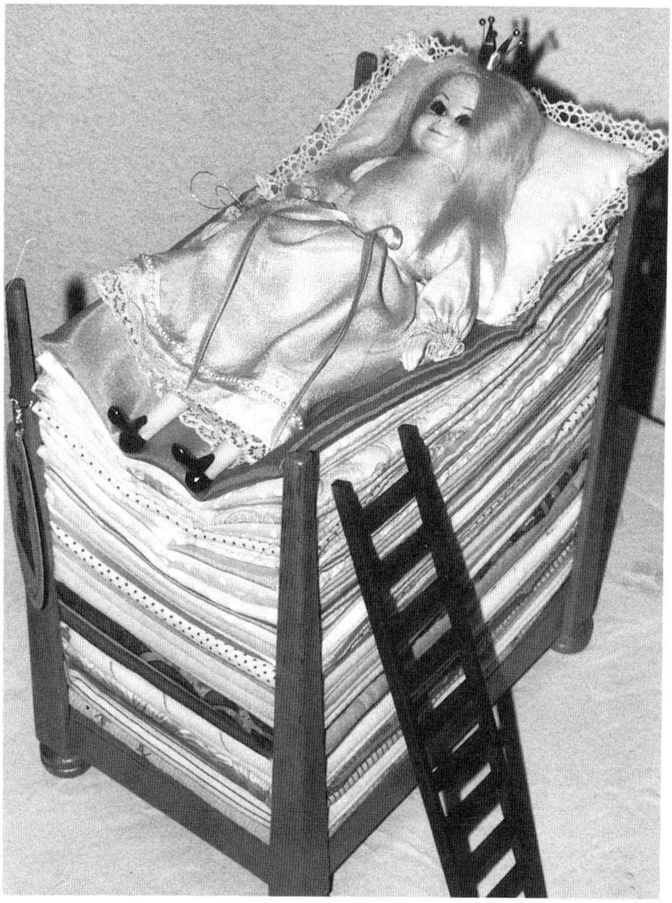

Ill. 33

(32) Hans Christian Andersen doll with artist. © Courtesy of Kirsten Gynther Eriksen.

(33) Doll illustrating fable: "The Princess and the Pea".

Ill. 34

Ill. 35

(34) King Henry VIII.
(35 & 36) Presidents A. Lincoln and G. Washington. © Courtesy of Kirsten Gynther Eriksen.

Ill. 36

Andersen seated before an audience of children. That group was commissioned for a museum in Japan.

Kirsten Gynther Eriksen's interest is not confined to Danish personalities. The English King Henry VIII (Ill.34) and his wives as well as the Danish historical and literary figures are readily recognized by their accurately styled costumes which reveal her well researched efforts. When we visited her in May 1989, she showed with modest pleasure her figures of George Washington and Abraham Lincoln (Ill.35; 36) "to you, the first Americans to see them".

One of her recent creations is an exquisite portrait doll of Her Majesty Queen Margrethe II (Frontispiece), a commissioned gift for the Queen's fiftieth birthday on Apr.16, 1990. This was Kirstin's first doll made in the likeness of a living person; she approached the challenge with intense concern for perfection. Many consultations meant many revisions in details. She wrote that even the piercing of the ears had to be very exact. Two specific Orders had to be included. The Order of the Elephant is the highest Order given by the Danish King. And the Government Order includes a portrait of her father, the late King Frederick IX, a portrait so small that a lighted magnifier is needed to see the details.

Kirsten Eriksen's August 1990 letter to us listed her midyear accomplishments in addition to the Queen Margrethe doll. The newest figures included Elizabeth I of England, Victor Borge, the second likeness of a living person, and the horse for the equestrian King Christian X, Queen Margrethe's grandfather. "The king's head, hands and legs", she wrote, "are out of the kiln now". Her next letter in October 1990 included the pictures of Victor Borge (Ill.37) and the completed Christian X on his horse (Ill.38).

The diligence with which she is working, highly encouraged by the success of the Queen doll and Queen Margrethe's personal thanks, means that she now surpasses her earlier production of one new creation each year.

At her home in Nyborg she keeps one of each of her dolls. She anticipates having a museum for the growing display. The Hans Christian Andersen doll remains one of her favorites. Details of her mastery of this talent will be discussed in Chapter 12, Today's Doll Artists and Artisans.

The Danish press, very interested in publicizing these wondrous dolls, has called her "a Danish Madame Taussaud". Susanne Kroyer's story [27], headlined "*Hun lever og ander for sine dukker*", translates: "She lives and breathes for her dolls."

Ill. 37

Ill. 38

(37) Victor Borge. © Courtesy of Kirsten Gynther Eriksen.

(38) King Christian X on horseback. © Courtesy of Kirsten Gynther Eriksen.

Chapter 3

Folk Dolls—Then and Now

Island Folk Costumes

Denmark's island communities have long preserved their identities and cultures in their traditions and their folk costumes. Very few or no changes are seen in those costumes on the islands of Amager, Fanø, Rømø, and Samsø.

The island of Amager, connected to the capitol city of Copenhagen by three bridges, is very well represented in the Danish National Museum. This island community became the garden of the city in the 16th century when the Danish king invited Dutch peasant folk to settle there and grow vegetables for the court.

Antique dolls in bridal costumes representing the island are exhibited in the Danish Folk Museum, a part of the National Museum in Copenhagen [28], and at Legoland A/S (Ill.39, a,b). Silk ribbons which were a distinctive part of a bridal costume are characterized by beautiful floral designs in shades of purple on black (Ill.40). A small black silk Amager bonnet (Ill.41) was photographed at the headquarters of the Danish Immigrant Museum in Elk Horn, Iowa in April 1990. This cap is shown on the lovely Northern European doll head (Ill.42) given to the museum in August 1990 by Mr. and Mrs. Madsen of DanDolls in Jutland, detailed in Ch.9.

In contrast with the rich dress of Legoland's Amager bride doll are the simplified styles in a series of Amager dolls made by Globus, a modern doll company in Horsholm. The firm was begun in 1961 in Nivå, where dolls were made in four sizes, from 5 in to 12 in (12.7 cm to 30.5 cm). Those dolls sold at the Copenhagen Sheraton Hotel gift shop in 1989 were 9 in (22.9 cm) tall. Globus dolls are well made and attractively packaged in transparent plastic carriers. They are available in Danish shops and are distributed in the U.S.A. by the Odin Sales Corp., West Islip, NY.

In the Globus firm's brochure five dolls represent the island of Amager. They include: 1. the everyday costume, a blue apron over a black dress and a black bonnet; 2. Amager *brud* (bride), a red apron and floral ribbon cascade over a black dress and a white lacy ruffled bonnet; 3. Amager *kirke* (church), a red and black lacy apron over a black dress and a red embroidered, black pointed cap; 4. Amager *kokke* (cook), a white apron over a black dress and a white trimmed black cap; and 19. Amager man in a

Ill. 39 a

Ill.39 b

Ill. 40

Ill. 42

Ill. 43

(39 a, b) German doll in Amager bridal costume, full length and close-up.

(40) Cascade of ribbons from a bridal costume of Amager. Photographed courtesy of June Samson, Director, Danish Immigrant Museum.

(41) Black silk cap of Amagar. Courtesy of June Sampson.

(42) DanDoll head wearing Amager cap. Photo courtesy of June Sampson.

(43) Three Amager folk dolls by Globus. Sketched by Eileen van Baren.

Ill. 41

Ill. 44

Ill. 45

Ill. 46

(44) Amager *Kirke* (church) doll by Gittes Dukketoj.
(45) Details of pointed bonnet of Amager church costume.
(46) Fanø Days festival showing costumes worn by adults and and children. Courtesy of Fanø Tourist Bureau.

black suit and hat. Details are shown in sketches of the dolls numbered in the brochure 3, 4, and 19 (Ill.43). The bridal costumes on both antique and modern dolls include the white trimmed bonnet, white lace at the edge of the sleeves, and the floral embroidered ribbons cascading from the waist.

The Amager *Kirke* in Sunday dress, one of 23 folk dolls given to the Danish Immigrant Museum, wears another popular interpretation of that Amager style (Ill.44, 45). The pointed bonnet gives emphasis to this costume. This series of dolls was made by *Gittes Dukketojsfabrik, Leck's Fabrikker*. The firm will be detailed following the dolls representing the other islands, Fanø, Rømø, and Samsø.

Variations in the head dress, shawls, and ribbons, and the functional costumes of the Amager folk, ca 1870, are described and illustrated in *Folkedragter* (Folk Costumes) [29]. The boys and girls were dressed almost alike until they were four or five years old. A long black sleeveless dress called a *klokke* was worn over a jacket or blouse with brocaded sleeves. A boy wore a small silver locket with his initials, a *solvore*, dangling from his cap in the position of an earring. Both boys and girls wore long dark aprons. This description gives another clue to what age the long dress was worn by a young boy, represented by the porcelain boy doll in Chapter 1.

A series of islands in the North Sea follows the coast from Friesland, the northern extremity of the Federal Republic of Germany, to Esbjerg on the Jutland coast. The southern group which is German is called the North Friesan Islands. Three very small offshore islands just north of them, Fanø and Rømø with a tiny dot of land between them, are in Danish waters. Fanø became a prosperous ship building center second only to Copenhagen in 1741, when the residents bought the island from the Danish king and became freeholders. Until 1900 ships of all sizes were built. The fresh sea air and attractive beaches, dunes, and heathlands have made the island a desirable holiday retreat for the Danes and foreign visitors for over a century. Festivals at which folk costumes are still worn are very popular. On the first weekend in July the *Fannikerdagene* (Fanø Days) celebrate the traditions of the town of Nordby. On the third Sunday in July the festival at Sonderho is highlighted by a wedding procession to the town square [30]. The traditional Fanø costumes and bridal dress are worn in the dances and festivities.

The women's dress of somber brown, grey, or other dark colors is brightened by plaid neck and head kerchiefs. Floral embroidered braid at the wrists of the long sleeves and down the left side of the bodice add colorful accents. The bodice, with a double row of buttons down the center front, usually matches the very full skirt wrapped almost to the center back. A fully pleated dark green skirt edged with a band of brighter green flares out from under the wrapped skirt, and a red

Ill. 47

Ill. 48

(47 and 48) Three dolls in Fanø costumes. Courtesy of Lisse Holm Nielsen and Alice Thorgaard.

Ill. 49

(49) Fanø doll bought at Legoland A/S.

petticoat swirls beneath both. The little girls who enter the dancing festivities wear similarly styled garments, their pleated skirts of red also edged with bright green. Dancing shows the colorful combinations of these costumes (Ill.46).

Three dolls in Fanø costumes (Ill.47, 48) were photographed at the October 1989 Euro Show at Legoland A/S. The small 9 in (22.9 cm) doll is made of hard plastic (Ill.49). She wears brown and white print trimmed with patterned black woven braid. The two rows of small buttons on the bodice, and the neck scarf and head scarf wrapped to a peak at the crown are typical details of Fanø folk dress on all three dolls.

The island of Rømø (Rømø translates 'sand bar', an island from a sand bar in the sea.) is represented very realistically in this presentation of dolls in folk costumes. The Petersen family owned *Kommandørgården*, a large farm at Toftum in northeast Rømø. The Jepsen family lived north of *Kommandørgården*. There were three Jepsen daughters; the eldest, Hansine, married Christian Petersen, owner of the farm, in 1901. Christian and Hansine lived at *Kommandørgården* all their lives. In 1950 the farm became a part of the Danish National Museum on the Island of Rømø, and a granddaughter of Christian and Hansine Jepsen Petersen became the director.

Hansine Jepsen Petersen wore her Rømø costume for a photograph by Lennart Larsen of the museum staff. The portrait and the costume are at *Kommandørgården*. Postcards sold there have been made from this photograph (Ill.50 a). An 1871 series of postcards showing Danish national folk costumes were colored and reproduced in 1990. Three of them show women wearing the costumes of Rømø, Fanø, and Amager (Ill.50 b, c, d), and are shown here with permission of the Northern Light Gallery in Copenhagen.

In 1950 the Danish Queen Ingrid and their three daughters began visiting the farm for a peaceful annual holiday. Today's Queen Margrethe II was one of the daughters; all became good friends of the Petersen family. Queen Ingrid's dressmaker had dressed a doll in a very accurate copy of the Rømø folk costume. The doll named "Bodil Marie" is a treasure of the third generation of the Jepsen family.

Bodil Marie is in the Charlottenlund home of Mrs. Gunhild Nissen Nielsen, daughter of the youngest sister, Meta Jepsen Nissen. Gunhild was photographed in 1989, holding the doll and a museum postcard of her Aunt Hansine Petersen (Ill.51).

A sketch of Bodil Marie details the costume with the names in Rømø dialect (Ill.52). The wrapped head scarf is called the *klud*. The blouse is the *bul*; the red woolen dress, fashioned like a jumper, is the *pi*. And the heavy gathered folds in the pi create the bustle effect called the *bunker*.

Ill. 50 a

Ill. 50 b

Ill. 50 c

Ill. 50 d

(50 a - d) Photographs of folk costumes from postcards:
a. Hansine Jepsen Petersen in Rømø costume. Lennart Larsen photo.
b. Girl in Rømø costume.
c. Girl in Fanø costume.
d. Girl in market day costume of Amager. Illustrations 50, b - d courtesy of Northern Light Gallery, Copenhagen.

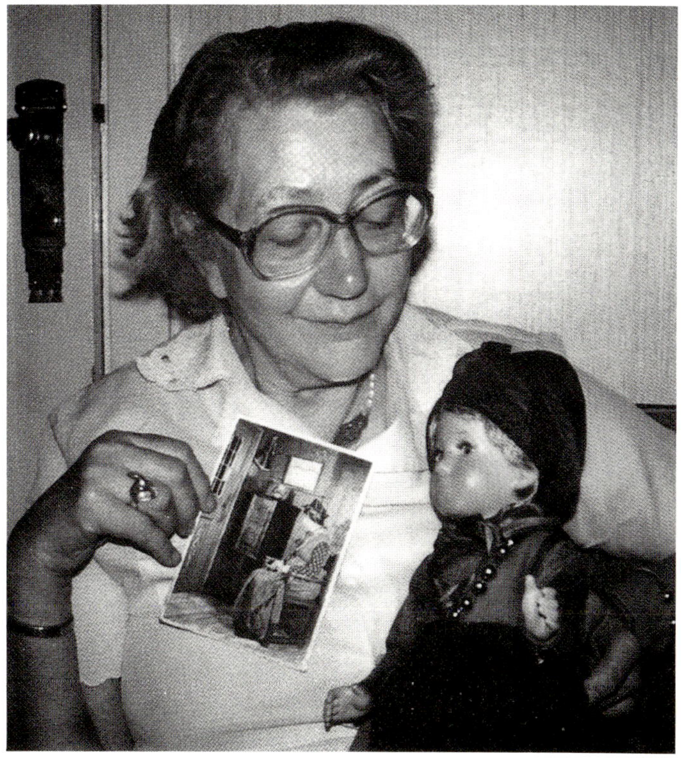

Ill. 51

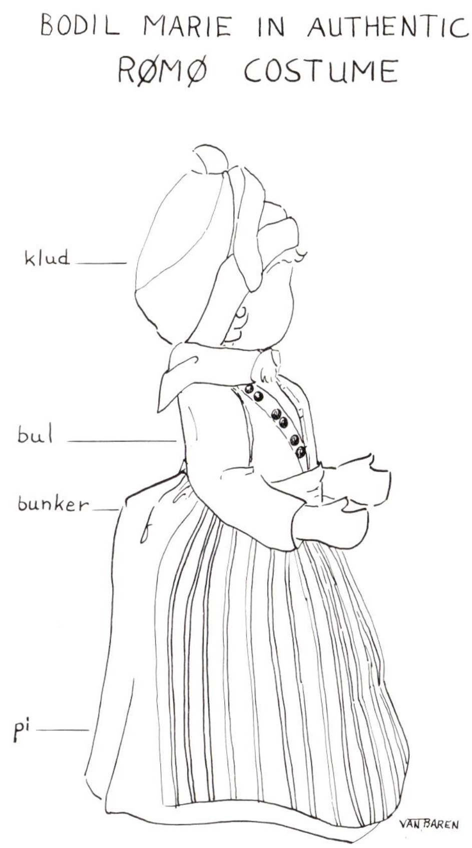

Ill. 52

(51) Gunhild Nielsen with photo of her Aunt Hansine and doll Bodil Marie.

(52) Bodil Marie in costume. Sketch by Eileen van Baren.

Gunhild Nielsen told us that "I remember from my childhood, in the 1920s, Old Jette who still wore *pi*." She showed us her family's complete Rømø costume, certainly a treasure, as is the doll Bodil Marie.

An earlier doll with a papier mâché head on a straw filled cloth body was found in the effects of the late Meta Nissen in 1979. No details are known about this doll, but her name became "Haldora", for the grandmother of Gunhild Nissen Nielsen. This doll came to Colorado with us in 1989.

Haldora is 16 in. (40.6 cm) tall, with unusually long legs. Those and the arms are secured to the body with rusted, flat metal discs on wires, giving shoulder and hip articulation. Her grey eyes, black eyeliner and lashes, and light brown eyebrows are beautifully detailed and as clear as when painted. Red tear duct dots, nostrils, and a red line between her pale rose colored lips are distinct. Her bruised nose reveals the grey papier mâché beneath the white primer and the dark peach colored paint. A thinning blonde mohair wig in two braids is secured by a piece of mohair through a hole in the center of the cardboard pate, a detail also seen in some early German papier mâché dolls. Under the wig are the incised marks "8/0//K".

This doll's Rømø costume needed some help. The white stockings and faded black-to-grey cotton print apron are intact. Sleeves and a dickey of the blouse match the apron; the bodice was made of a pink print (which doesn't show), and all was carefully lined with heavy white muslin. Sleeves were very worn and needed stabilizing. A navy and green cotton plaid has been used to replace the missing head kerchief and the other at her neck. Old white cotton pantaloons and petticoat from clothing of china head dolls are added. The largest holes in the red woolen petticoat and dress have been stabilized by basting red wool patches under them. The green silk binding of the armseyes and deep V neckline was replaced only where completely worn off the left front. Tiny silver buttons at the neckline of the *pi* and the silver buckle fastening the apron band are original; they are necessary details of the Rømø folk costume. The doll (Ill.53, 54) and her clothes (Ill.55) are shown as acquired; she now wears the renovated costume (Ill.56).

The island Samsø is located east of Jutland and almost directly north of Odense, the large city on Funin. A bisque reproduction doll at the Euro Show at Legoland A/S (Ill.57) wore a dark blue dress with long sleeves, and a striped apron with its plaid belt tied in front. A white lace edged cap was decorated with a blue bow. This doll was said to be dressed in the folk costume of Samsø. The Samsø doll by Globus (Ill.58) wears a black dress and a long full apron of white lace. The lace trimmed bonnet is tied with pink ribbons; matching ribbons cascade from the center front of the apron's belt. The doll shown from a 1980 brochure [31] is dressed exactly as the one in the illustration from Globus in late 1989. The Globus costumes are a bit more elaborate; perhaps the one at the show was an everyday dress?

One National Costume; Inland Folk Dress

A beautiful Globus doll, *Mor Danmark* (Mother Denmark), wears the Danish national colors, red and white (Ill.59). The white lace bonnet and lace apron over a red skirt are complemented by a little red bolero jacket over a white blouse. Not a regional folk doll, she is a representative of the entire country of Denmark; might she be considered a national doll? (I will dress one of our several small Danish celluloid dolls as *Mor Danmark*.)

Other attractive Globus dolls in folk costumes represent the cities of Odense and Copenhagen. A man and lady are dressed as older residents of the capitol city. Another firm, Formby in Copenhagen, made a doll representing the city (Ill.60) whose costume is very similar to the Globus lady doll's dress. The label at the hemline of the skirt is marked "Gudrun//Formby//Danish// Design//Copenhagen Denmark". This Formby doll was a 1983 gift to a friend from the Danish-American town of Solvang, California.

A colorful brochure illustrating seventy five "DANISH DOLLS IN ORIGINAL NATIONAL COSTUMES" was sent to me from *Leck's Fabrikker, Gittes Dukketojsfabrik* in Praestø. For 17 years this firm has costumed dolls which they import from another European country. The dolls of two sizes, about 6 1/2 in (17 cm) and 11 in (28 cm), are dressed under the supervision of Mrs. Kirsten Leck Andersen. She attempts to keep the simplified costumes as realistic as possible, and continues to make needed subtle changes.

A bonanza of 23 eleven inch dolls made by this Praestø firm was given to the Danish Immigrant Museum by Mr. William Travis of Munster, Indiana, a loving memorial to his late wife, Erma Travis. Some of the dolls photographed there (Ill.61 a - f) are those representing: a. Arhus/Randers, b. Rebild, c. Praestø, d. Solvang, e. *Nordsjaelland* (North Zealand), and f. *Sonderjylland* (South Jutland). The South Jutland costume of unusual colors accurately represents that shown at the Danish Folk Museum. Many of the men there were sea captains with

Ill. 53

Ill. 54

Ill. 57

Ill. 58

Ill. 55

Ill. 59

Ill. 60

Ill. 56

(53, 54) The doll Haldora as found: undressed; in parts of Rømø costume.

(55) Haldora's original clothes.

(56) Haldora in renovated Rømø costume.

(57) Reproduction bisque head doll in Samsø costume. Courtesy of Torres Housewares.

(58) Samsø doll made by Globus. Courtesy of U. Gustavsen.

(59) *Mor Danmark* doll by Globus. Courtesy of U. Gustavsen.

(60) Formby Doll of Copenhagen, courtesy of Betty Newland.

Ill. 61, a

Ill. 61, b

Ill. 61, c

Ill. 61, d

Ill. 61, e

Ill. 61, f

good incomes; they were proud people who liked to dress well [32].

Studying the brochure helped identify small dolls in our collection which should be included here. A pair sent ten years ago from Charlottenlund are the girl *Lokken/Garder* (guard of Lokken) and the boy *Garder* (guard) of Amalienborg Palace (Ill.62), both attractive in red and white. Another doll from the Praestø firm is the girl guard of *Helsingor* (Danish name for Elsinore). Her blue and white costume (Ill.63) is styled like that of the girl guard of Lokken.

Greenland Folk Costumes

A picture in the brochure from Mrs. Kirsten Leck Andersen identified a doll from *Grønland* (Greenland), 8 1/2 in (21.6 cm), with its elaborate coiffure, and braid and lace trims (Ill.64). The Danes' interest in the Greenland costume is shown in the annual *Denmark Cavalcade* books of photographs. On the cover of the year 1952 [33], the three Princesses, Margrethe, Anne Marie, and Benedikte, are wearing the Greenland costumes presented during the summer visit there. The Greenland festival costume given to Queen Ingrid shows vertical braid from the toe of the boot to the knee, where it meets more braid in a V (as on the doll's boots). The tall red boots are "made of sealskin, the so-called '*kamiker*'". In the 1965 Cavalcade [34], the Queen is shown strolling with a score of Greenland women, all wearing the colorful costumes. Typical trimming on their white boots and on those of the princesses are embroidered panels near the tops, and deep cuffs of lace edged floral embroidery. The wide standing collar of black is worn with all those costumes, and with the Queen's festival costume.

Although they are not really dolls, a series of porcelain figurines in folk costumes is mentioned here. They were made for the occasion of a Royal anniversary in 1923 by the Royal Copenhagen firm. The sculptor was Chr. Thompsen [35]. The boy and girl of Greenland are readily recognized.

(61 a - f) Dolls at Danish National Immigrant Museum:
a. Aarhus/Randers
b. Rebild
c. Praestø
d. Solvang
e. North Zealand
f. South Jutland.

Ill. 62

Ill. 63

Ill. 64

(62) Lokken/Garder and Garder.
(63) Girl Guard of Elsinore.
(64) Greenland doll.

Artist Dolls in Folk Costumes

Dolls dressed in regional folk costumes have been made by doll artists in Denmark. A beautiful doll with a wooden head (Ill.65) made by Mrs. Pamela Nedergaard in 1983 represents her city, Odense. Pamela described the Odense dress in the brochure included with the dolls (36):

> The Odense costume worn by the young ladies of the town was of bright red, with touches of lace and blue ribbons. Her bodice is worn over her apron, and the lace earflaps which trim the little red silk bonnet add a special and "local" touch to the Odense costume.

One of Pamela's little Odense dolls has been presented to Her Majesty Queen Margretha II.

A colorful pair of dolls in the folk costume of Salling is picturesque, especially the girl's white cap with wide flaring brims stiffly starched (Ill.66). The Salling folk costume is widely known through the drawings and paintings of Christian Dalsgaard (1824-1907) [37]. The doll pair was made and accurately dressed by another talented artist and needle worker, Mrs. Aase Breivik of Gentofte. Her specialty is folk dolls, and her inspiration is *Folkedragter*, the well illustrated book of folk costumes [38] by Minna Kragelund, formerly on the Danish National Museum staff. When Aase Brevik moved to Norway it was not possible to photograph or document more of her representative folk dolls.

Pamela Nedergaard and Aase Breivik will be profiled in Chapter 12 with other doll artists we have met and interviewed in Denmark.

The last folk costume shown (Ill.67) is worn by the Dam® Troll as another *Garder*. This amusing fellow will introduce Chapter 4, all about his creator.

(65) Artist doll in Odense folk costume.
(66) Pair of Salling dolls. Courtesy of Aase Breivik.

Ill. 65

Ill. 66

Ill. 67

(67) Dam® Troll in *Garder* costume. Courtesy of Niels Dam.

Chapter 4

The Dam Trolls

The Troll Legend; The Troll Family

Common in the folklore of the Scandinavian countries are the Wee Folk, the Trolls and the elves called *Nisser*, which are more than 200 years old. The very popular and collectible Danish trolls, sold worldwide and frequently copied, are the DAM® TROLLS made in Gjøl in northern Jutland.

The story of the Trolls, the "Dam Things from Denmark", is the story of the three generation Dam family. Mr. Thomas Dam, 75 years old when we visited the firm in May 1989, was a fisherman in the Lim Fjord at an early age. He became a baker in the village of Gjøl. When his customers could no longer pay for bread, "he went broke with the bakery, or there never would have been a troll factory" [39]. He turned to his hobby, his skill of woodcarving, and was soon selling ten to twenty pieces to the wealthy burghers in Aalborg, the city across the fjord. By 1955 he had established a clientele who competed for his products. He had carved his first little troll in 1951; by 1955 his wife and sister were making trolls in their home as toys for the children. Factory production was begun in 1959, and more than five million trolls were made in the next thirty years.

Mr. Thomas Dam continued making new designs, new creations, until his death, Nov. 12, 1989. Their son Niels, with the help of their daughter Lajla, makes the molds and begins the production of each new line. The artistic talents of the family are evident in these remarkably successful toys. The grandson Jeppe is in charge of sales and distribution. The company employs more than thirty people; the daily output in 1989 was at least 3,000 items.

Very innovative and very business minded, the Dams have tried numerous designs, most of which are extremely successful. A troll dressed in red and white (Ill.68) became Troligan, coined from "Roligan", from "Hooligan", meaning "one who takes things easy", according to Niels Dam. This mascot was created for the occasion of the Danish football team playing for the European championship in France in 1984. Within three months 10,000 Troligan mascots were sold. But when the team lost the game the Danes lost interest, and no more were sold. Another character which promised success was Quark, a star of the 1986 Danish film *Valhalla*.

Ill. 68

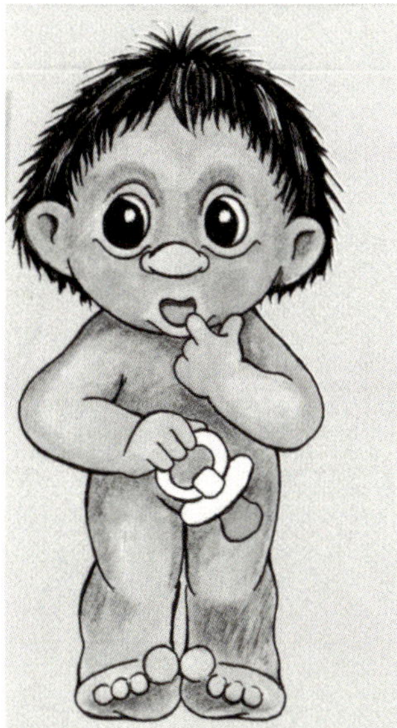

Ill. 69 a

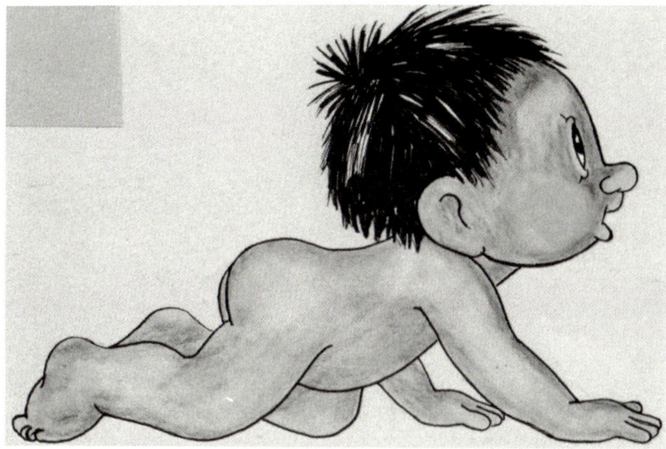

Ill. 69 b

Ill. 70

Ill. 72

Ill.71

All photographs made in Gjøl and from printed material obtained there, reproduced with permission of Mr. Niels Dam, with thanks.

(68) Troligan created as football mascot.

(69 a, b) Tiny Troll, sketches from prospectus. ®DAM.

(70) Wood carving by Mr. Thomas Dam: Troll Couple.

(71) Newborn Trolls, ®DAM. Poster reproduced courtesy of Niels Dam.

(72) Grandpa and Grandma Stinne, Gwenda Ward and D. Luckey collections.

Ill. 73

Ill. 74

Ill. 76

Ill. 75

(73) Representative Trolls on cover of 1989 prospectus. ®DAM.
(74) Grandma Stinne, with picture of Nisser. ®DAM.
(75, 76) Clowns, current items, in ruffled clown suits, in body suits. Courtesy BR Toys.
(77) Souvenir trolls in *Danmark* tee shirts. ®DAM.

Ill. 77

This film, with a Disney-like quality, went broke a year later, and an entire cast of potential Dam Things was abandoned.

The continuous flow of ideas has kept the Troll Family and the company alive and well. The 2 3/4 in. (7 cm) Tiny Troll (Ill.69), child of the peasant pair in the delightful wood carving (Ill.70) hanging behind Niels Dam's desk, is cleverly presented as the Newborn in the 1989 poster (Ill.71).

Quoted with permission of Niels dam, the Troll family history is narrated in the 1988 Prospectus, *Portrait of a Successful Family* [40]:

> Once upon a time ... that's the way all good fairy tales start, and the story of the Lucky Trolls from Gjøl is no exception. Many years ago — so many in fact, that no one knows exactly how many — a kind, old troll named Uras lived on the small, forested island of Gjøl in the Lim Fjord in Denmark. Uras lived in a cave with his very large family. There wasn't much room, and as the family continued to grow, Uras decided to build a new, bigger cave. And to make this work a little easier he tried to cast an ancient, secret magic spell to make the trolls bigger and stronger.
>
> Now as most people know, trolls aren't especially good at casting spells and using magic formulas, and sure enough, when he tried the formula, the trolls began to shrink. Uras tried again and again, but it was no use. They just got smaller asnd smaller, and soon the troll children were so small that they could hide from the bigger Uras, and they scurried out of the cave and ran away. That is why today you can find these tiny, happy little beings all over the world...Tiny Troll, a sweet, happy little fellow, and —as far as we know — the youngest member of the Troll family, lives on the beautiful island in the Lim Fjord, (with) ...all his old Troll family relations.

This family tree is extensive. Grandpa and Grandma Stinne (Ill.72), pleasingly wrinkled, are 14 1/4 in (36 cm) tall. Papa and Mama, 9 13/16 in (25 cm), Hanna and Hans who are sister and brother, 8 5/8 in (22 cm), an Aunt Tulle, 9 13/16 in (25 cm), are the rest of his immediate family. Numerous friends as well as kinfolk (Ill.73) make up more of the population. The Troll Kids from the great forest of Rold live in a cave with their pet pigs, Oink and Grunt. The kids from the Black Forest have a tree house, and more Cousins visit from the Fos Valley. Brother John at the Monastary cares for sick oldsters and orphans. Vena and Simon, the anatomically correct girl and boy, live with him. Their talented friends include Jane who wears a leotard and leg warmers for gymnastics, Thai, a student of karate proudly wearing his black belt, and Skipper who sails an old ship in the fjord. The tallest pair, 18 7/8 in

Ill. 78

Ill. 79

(78) Five animal banks.
(79) Mr. Niels Dam with seal bank.

Ill. 80

Ill. 81

(80) "Porky Pen" with turtle bank inspiration.
(81) "Writing Trolls". ®DAM.

(45 cm), are white haired and distinctly nationalistic in red and white costumes. Of course there are many Christmas Nisser in red, white, and green; they will be encountered in Chapter 5.

New additions in 1990 included the happy looking Lucky Trolls who will appear in Chapter 12. A photograph of Grandma Stinne with the picture of the group of Nisser of all sizes shows the change of these oldies into the *Julemand*, Santa, and Mrs.Claus (Ill.74). Clowns in ruffled suits and pointed caps (Ill.75) have typical red spots on their noses and cheeks; other clowns wearing long body suits have wild tufts of hair over each of their ears (Ill.76). Other trolls have been costumed by the E. F. S. Marketing Assn., 164 Central Ave., Farmingdale, N.Y. Their tourist trolls in Hawaiian dress, and boys and girls wearing tee shirts printed "I love (as a heart) New York" are popular. The current slogan on the trolls, "Adopt a Norphin", originated with that U.S. distributor, Niels told us.

Souvenir Trolls sold in Denmark wear tee shirts printed with the Dam Troll trademark and "*Danmark*" (Ill.77). Others had been dressed in tee shirts with names of banks such as *NORDJYLLAND* (North Jutland) as premiums ca 1953. Three years later the banks offered them for sale and finally, with little demand for them, their manufacture was discontinued. Dam did not give up on banks. A new series of animals as coin banks (Ill.78) made in 1989 is now marketed. Very appealing in this series are the blue eyed whale, and the seal shown by Niels Dam (Ill.79). The bank made as a smiling turtle with hypnotic eyes was the inspiration for another promising new item, the Porky Pen (Ill.80). A customer had asked the Dams for an animal — a hedgehog or a porcupine — to hold felt tip pens. This is an excellent product; the caps of the pens so easily swallowed by children may be discarded because each of the 24 pens stands, felt tip down, in a hole in Porky's back. Nontoxic ink is used in the colored pens. This kit was priced from 85 to 100 D. Kr., $12 to $14, in May 1989; the safety factor will surely sell it. On the bottom the porcupine is marked "MADE IN Denmark// © Dam Nitusch® 1986".

Another simple and very inexpensive design is the tiniest Writing Troll (Ill.81) who fits atop a pencil and makes chewing the eraser impossible. A child is smugly told that the troll will "keep an eye on you and be sure you write nicely" [41].

Production of the trolls begins in the old building in the town of Gjøl, where the rotation machines are located. The figures are removed from the hot copper molds while the vinyl is still warm. All the finishing is done in the large airy building outside the town, where they are painted and hair is added (Ill.82). Real sheep wool from Iceland, long and thick, used for the wigs of the girl trolls because synthetic furs are not long enough, was shown to us there. They are dressed; the eyes are added last. All the trolls are made of the finest non-toxic vinyl, another safety feature of the Dam production.

Ill. 82

Ill. 85, a

Ill. 85, b

Ill. 83

Ill. 84, a

Ill. 84, b

(82) Hair being applied to troll.
(83) Bridal trolls at 1968 Silver Anniversary. Don Loomis photo.
(84 a) Bridal pair, 1989. ®DAM.
(84 b) The 1992 3 1/2 in (8.8 cm) Bridal Party Norfins, courtesy of Eva Stark, E.F.S. Mkt. Assoc. Inc.

Ill. 85, c

Ill. 85, d

Ill. 85, e

Ill. 85, f

On the older building in the village a sign proclaims: "*GJØLTROLDEN DANMARKS NATIONALTROLD*" which translates "Gjøl Trolls, Denmark's National Trolls". Denmark and Mr. Thomas Dam are interchangeable symbols of the father of the Trolls, now made in many other countries. Trolls are internationally known and are enjoyed by children and by collectors everywhere.

Trolls in Doll Literature, in Collections

In more than three decades of their manufacture begun in 1959 the trolls have become very popular collectibles. They have been made in other Scandinavian countries, in Japan, Hong Kong, and Ireland. Canada's Regal Doll Company and three prominent U.S. doll manufacturers, Uneeda, Ideal, and Russ have added to the variety.

Trolls have appeared in doll books since 1976, when Patricia Smith first included two Uneeda Trolls in her *Modern Collector's Dolls, Third Series* [42]. Mrs. Smith's *Fourth Series* [43] shows five trolls. A felt dressed 5 1/2 in (14 cm) troll, colorful with chartreuse hair and blue eyes, is marked "Dam Thing" in a circle on its back. A pair, 3 in (7.6 cm), is unmarked, but facial features and bodies are very much like Dam's trolls. In Smith's *Fifth Series*, 1984 [44], the trolls appear on four pages. A 12 in (30.5 cm) very fat Santa, with green eyes and bearded and wigged in white, is marked "DAM THINGS ESTABLISHMENT// 1964" on his foot. Small animal trolls, 3 in (7.6 cm) each, unmarked but looking like documented Dam products, include a cow and a donkey. The 18 in (45.7 cm) black troll, wearing only plastic baby pants, is marked "DAM THINGS EST. 1964" on his head.

In a 1987 book of black dolls [45], a 9 1/2 in (24.2 cm) dark brown troll has long brown wool hair in braids, and brown glassene eyes. This one is marked in a circle, "Thomas Dam//Made in Denmark". Dark skinned trolls were made in fewer numbers; today they are marketed as Norfins.

(85 a, b, c, d) Three Dam troll kids and a Viking, postcard prototypes. 1966. ®DAM.
(85 e, f) Two animals, postcard prototypes. 1966. ®DAM.

Within the same fifteen years doll price guides have included trolls. Bill Schroeder's 1976 value guide [46] shows red-orange and green haired creatures, Hong Kong copies. Patricia Smith's 1980 *Series II* value guide [47] includes three Thomas Dam trolls, an Ideal Scandia House pair, and a troll monkey.

With the trolls' appearance in recent doll magazines, these wee creatures have certainly found their niche in dolldom. One author [48] shows over 15 Dam Trolls from the early 1960s. Two of the current models made by Dam are shown with the Norfin labels. That author's later article [49] contains excellent pictures of some of those same trolls in her personal collection; variations in the skin tones and especially the wild hair colors are remarkable. Patterns for clothes for the popular 9 3/4 in (24.9 cm) Dam troll have been given in a crafts magazine. The designer has advised that the distributor of the Norfins, E.F.S. Marketing Associates, "is developing a troll fan club, too!" [50].

The favorite toys of our Danish friends' children when we visited them in 1961 and 1962 were their tiny teddy bears and small trolls. Their daughter Birgitte later wrote a story including the two trolls, prominant in their play acting. The pair of trolls had been redressed in hand knitted garments. One had lost an eye, the glassene type used at the Dam factory, and wigs of both showed much wear.

Our daughters collected the DAM THINGS in the 1960s. Unfortunately none of them remain for photographing, but the small one with chartreuse hair and another wigged in pink are well remembered. In 1968 a bridal pair of DAM THINGS (Ill.83) was sneaked into the decor for our Silver Anniversary celebration by our collegian daughter-hostesses. The Dam trolls have routinely included the brides and grooms. Bridal pairs with variation in the costumes were seen in the BR Toy Shops in Copenhagen in May 1989, and appear in the recent catalogues of Dam and E.F.S. (Ill.84 a, b).

The 1966 series of photographs for postcards was not printed for marketing. Four trolls, a cow, and an elephant are illustrated (Ill.85 a-f). A small horse and a mouse were also made about that time; the pigs are delightful 1989 pets. The dated pictures given to us by Niels Dam will help collectors identify their Dam trolls which are not always marked with the company's name. The unmarked 2 5/8 in (6.7 cm) cow and the 12 in (30.5 cm) giraffe, incised "THOMAS DAM//MADE IN DENMARK" (Ill.86), were found in the flea market in Copenhagen in 1979.

One of the five banks we got from Mr. Dam in 1989 was not marked. This is not unusual nor intentional in doll manufacturing.

Trolls of 1964, 1978, and the current Norfins are shown in a series of photographs (Ill.87-90). A recent addition is the anatomically correct girl and boy pair of

Ill. 86

Ill. 87 a-c

(86) Cow and giraffe, older toys.

(87 a-c) Early Dam trolls. Colleen Hart collection.

a. 5 1/2 in (14 cm) troll marked "Thomas Dam (in oval)//U.S. Foreign Pats. Pending".

b. 2 1/2 in (6.4 cm) to hairline of smallest troll; unmarked, it is a Dam Troll.

c. 7 3/4 in (19.8 cm) troll marked: "1961/190918// Made in Denmark (in oval//U.S.A. Foreign Pats.Pending".

Ill. 88

(88) Small mouse, 2 in (5.1 cm); verified as Dam product by Niels Dam.

Norfins, continuing the sexual revolution in doll design begun in 1967 in France.

"The company Dam Things has changed its name to: Troll Company ApS", advised Niels Dam on September 26, 1991.

That old troll Uras would be surprised to realize just what he started so long ago. He would probably agree with Jacqueline Wilson that "Tracking the troll can be a challenge .... but the fun is legendary!" [51].

Ill. 89 a

Ill. 89 b

Ill. 90 a, b

(89 a, b) a. Troll, 8 1/2 in (21.6 cm), new 1979, with his troll house carton. Mark on left foot: "Thomas Dam//© "A Danish Good Luck Troll".

b. Troll house carton: 6 1/2 in (16.5 cm) X 4 3/4 in (12.2 cm) X 4 5/8 in (11.8 cm). Text: "I'M a DANISH GOOD LUCK TROLL...." // "DAM THINGS FROM DENMARK ApS". Luckey collection.

(90 a, b) Two NORFINS.

a. Newborn, 3 3/4 in (9.6 cm). Marks on back: "©DAM 198_". On plastic box: "Newborn/ TM//norfin...//the baby of the bunch" and "Imported by//EFS Marketing Assoc.//Farmingdale, NY//a Thomas DAM Design//© 1986/Model #510".

b. Black Baby boy, anatomically correct. Ht. 11 1/2 in (29.2 cm). Marks, on back: "1972//© ". On label, front: "ADOPT A//NORFIN TM".On label, back: "STYLE #3301//'BABY BOY'. "Imported Exclusively By//E.F.S. MARKETING ASSOCIATES, INC.//Farmingdale, New York//Made in Denmark by//Dam Things ApS". Luckey Collection.

Chapter 5

Danish Nisser, The Christmas Elves —Then and Now

Who are the little people dressed in grey and red, who inhabit secretive places such as the teacups in the china cupboard, and observe the family during the weeks before Christmas? They are called *Nisser*; the singular is *Nisse*.

Several Danish friends have told us that the Nisser, the Wee Folk who had lived in barns and brought luck to the farmers for over 200 years, were originally from Norway's folklore. A study of historical maps [52] shows that from 1648 the Kingdom of Denmark and Norway consisted of all of Norway, the southern tip of Sweden, and the three parts of today's Denmark. By 1763 each country had its own identity; by 1914 Denmark no longer claimed the southern part of Jutland called Schleswig. Today's Denmark had been established by that time. Might one assume that the Norwegian Nisser are indigenous to all that land area, and are therefore as Danish as they are Norwegian?

The helpful Little Folk traditionally favored the farmers by protecting the animals and the crops just as long as they got their daily ration of rice flavored with butter. In later years they became the symbol of Christmas time in Denmark. The *Julemand* (Christmas man, Santa) brings gifts; the Nisser's role is to insure a happy Christmas for everyone, and plentiful harvests the following year. The Nisser play a prominent role in decorations, murals, and stories.

The *Nissetroldene* (Ill.91) made at the Dam Troll factory at Gjøl, Jutland, in sizes from 15 in to 2 3/4 in (7 cm to 38 cm) are popular everywhere. These white haired and sometimes bearded trolls become instant and loveable Nisser when dressed in red pointed caps and robes or jackets. The Tiny Trolls are amusing, wearing only red caps over their long white hair or occasionally a white felt tunic. The tallest ones become the Julemand and his spouse.

Mr. Dam had made large statues of the Nisser of wood and metal and dressed them in real clothing. One commissioned by a department store in Sweden is a seated

Ill. 91

Ill. 92

Ill. 93

Ill. 94

Ill. 95

(91) *Nissetroldene*, Nisser created by costuming white haired DAM® Trolls, courtesy of Mr. Niels Dam, Gjøl.

(92) Nisse statue of painted gypsum; photographed courtesy of Mr. Bent Borgesen; sketch by van Baren.

(93) Old Nisse doll; photographed courtesy of *Dukke* Ella; sketch by van Baren

(94) Old Nisse with Greenland folk doll; photographed courtesy of *Messer's Dukkemuseum*, Ebeltoft; sketch by van Baren.

(95) Old faded Nisse; photographed courtesy of Toy Museum, Copenhagen; sketch by van Baren.

figure so large that his raised arm reaches the floor above. A child can be seated on a sled which enters that hand, glides through the body of the Nisse and exits through a foot on the main floor.

On our numerous extensive doll hunts in Denmark we have found interesting old Nisser, often faded but always wearing pointed caps. A very old gypsum statue of a Nisse's head and shoulders, about 20 in (50.8 cm) high, has a warm and kindly expression (Ill.92). This one could not have been an outdoor figure, the Copenhagen antiques dealer Mr.Borgesen said, because the gypsum would have weathered badly. It might have been a family's traditional piece, handed down from generation to generation until it appeared in the antique shop. Another Nisse, full figured and dressed as a doll in green jacket, brown trousers, red cap and scarf (Ill.93) is 15 in (38 cm) tall. This one was photographed at a Copenhagen shop called *Dukke* Ella. (*Dukke* is doll.)

At *Messer's Dukkemuseum* in Ebeltoft, a picturesque resort village on the Jutland coast north of Aarhus, a very old and faded Nisse stands next to a papier mâché Greenland doll in folk costume (Ill.94). A larger one of similar structure and wearing a faded red suit (Ill.95) is in the *Legetoj* (Toy) Museum in Copenhagen. Both of these oldies wear pointed red caps and thin black belts.

A felt Nisse doll by Steiff, ca 1878, is characterized by the center front seam in his face, and his tall pointed cap. This one is 15 1/4 in (40 cm) tall. It is illustrated in a Danish book about old toys [53].

A 10 in (25.4 cm) cloth bodied Nisse with a celluloid face mask and a white fur cloth beard was found in Copenhagen in May 1989. He is a stuffed cotton doll, the red suit becoming his body and limbs (Ill.96 a, b). A narrow black belt continues over his shoulder and holds a small black bag against his back. The function of the little bag, noted on two of the previous figures, has not been determined; it is obviously too small to be considered a Santa's toy bag. This funny fellow was thought by some to be a very young Julemand. The seller called him a Nisse and dated him late 1930s. Another doll, beardless, capless, faceless, and very faded, but of the same size and construction, was seen at an antique shop soon after we had bought the first one, without which that second one would probably have been unrecognizable.

Another Nisse with a celluloid head on a cloth body (Ill.97) was found later in 1989. Facial details are brightly painted. Without his pointed cap this one is about 8 in (20 cm) tall. He has lost his boots but wears the classic grey trousers over the simple red body and the pointed red cap. Similarities to the other celluloid Nisse suggest that he may have been a product of the late 1930s also.

At Casa Victoria Antiques in Copenhagen in 1984 we found a series of reproductions made from very old molds. Each December this painted plaster group, complete with the bowl of buttered rice, graces our mantle. All of these figurines are bearded; all wear red and grey, the classic

Ill. 96 a

Ill. 96 b

Ill. 97

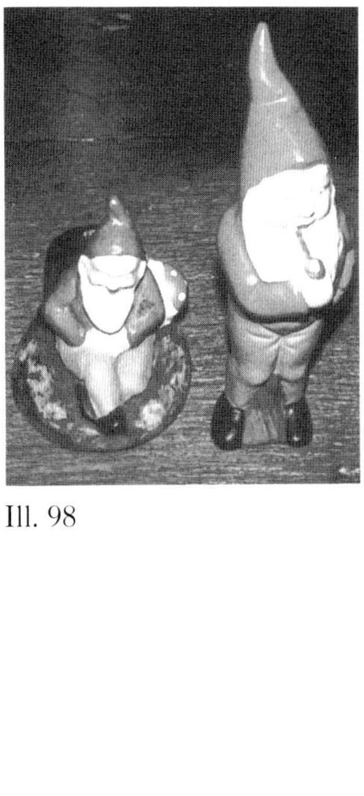

Ill. 98

(96, a, b) Cloth bodied Nisse, celluloid face mask, ca 1930;
a. front
b. back view.
(97) Cloth bodied Nisse, celluloid head.
(98) Nisser reproductions; courtesy of *Dukke* Ella.

Ill. 99 a

Ill. 99 b

Ill. 100

Ill. 102

Ill. 101

Ill. 103

(99 a, b) Nisser figures a. with candlestick, b. with pipe, photographed at Israels Plads flea market, 1989. Sketched by van Baren.

(100) Modern stonewear Nisse candlestick; photographed courtesy of *Illums Bolighus*, 1989. Sketched by van Baren.

(101) Hard plastic Nisse girl doll made in Italy.

(102) Wooden head youthful couple made by Pamela Nedergaard of Odense.

(103) Wooden head elderly couple made by Pamela Nedergaard.

Ill. 104

Ill. 105

Ill. 106

Ill. 107

(104-107) Needle sculpted Nisser dolls and animals illustrating legend of Grandpa and the Mice, by Pamela Nedergaard.

colors of the wee gnomes. The tallest is 2 1/2 in (6.3 cm). Some reproduction trolls photographed at *Dukke* Ella in 1989 (Ill.98) stand 8 in (15 cm) tall. She dated them ca 1975 and later. They wear the classic grey trousers, red jackets and caps; some of these bearded creatures smoke their pipes.

Small, old, and reproduced Nisser figurines were sold at the Israels Plads flea market in Copenhagen in May 1989 (Ill.99). Two which are 2 in (5.1 cm) tall hold candlesticks, and one a mere 1 1/4 in (3.8 cm) has a pipe. The modern Nisse candle holder (Ill.100) was photographed in 1989 at *Illums Bolighus*, an exclusive shop on the *Stroget*, Copenhagen's popular Walking Street. That stylized Nisse made of stonewear has unglazed hands and face, a shiny red glazed cap and a white glazed robe. His rounded mouth might be singing a Christmas song.

An 8 in (20.2 cm) hard plastic doll (Ill.101) sent from Denmark in 1974 was made in Italy. This articulated Nisse girl wears a 2 in (5.1 cm) tall red knitted cap, a red blouse and grey skirt of felt, and wooden clogs. This is the first modern Nisse doll we have documented; similar vinyl plastic Nisser are now imported from Taiwan and Hong Kong.

A variety of Nisser of numerous materials and sizes was photographed with permission in May 1989 at the home of Pamela and Ove Nedergaard in Odense, Funin. Pamela is a talented doll artist who has developed many styles of Nisser. Two Nisser couples with painted wooden heads were made in 1982. One seated pair, a girl of 8 in (20.2 cm) and a boy 9 1/2 in (24 cm), are youthful and winsome (Ill.102). Their elders, 12 1/2 in (31.8 cm), are bespectacled, and heavily bearded "Grandpa" smokes a pipe (Ill.103). All four dolls are charmingly dressed. Pamela costumes her Nisser in wooden or straw clogs over heavy knitted socks, long skirts, and shawls on the women, and knickers and sleeveless jackets over shirts on the men, creating unique and individualized figures. Her present Nisser dolls are made in soft sculpture, also called needle sculpture. She said she first creates the head; its size determines the body proportions. " The finished head also determines the sex of the Nisse; sometimes an added beard determines a sex change".

Each of her needle sculpted dolls is different and has its own distinct personality. A favorite group, her first oldsters, soft sculpted, illustrates one of the Nisser legends Pamela demonstrated (Ill.104 - 107): "Grandpa" is seated on his birch log wanting to eat his rice. Because the mice (of grey felt, 2 5/8 in [6.5 cm] tall) have become so bothersome, he summons the farm cat (crocheted of black yarn, 3 3/8 in [8.5 cm] tall) to drive them off. After the mice are frightened away, aproned "Grandma" joins him on the birch log and resumes her knitting.

One very impish fellow with big pointed ears and tousled white hair is clean shaven; he, too, sits jauntily on a birch log and smokes his white clay pipe (Ill.108). Pamela's recent creation she named "Mr. Happy". He sits

Ill. 108

(108) Clean shaven needle sculpted doll on log by P. Nedergaard.

Ill. 109

(109) Pair of Nisser, "Mr. Happy" and "Mr. Grumpy", in red chairs, by Pamela Nedergaard.

on his own red chair next to "Mr. Grumpy" in his red rocker. And both smoke their brown clay pipes (Ill.109). She is making a similar "Mr. Lucky" to send to us. Other Nisser in this group may be seated on red sleds, or on the birch logs or stumps.

Pamela makes her Nisser to order; she said she has no desire to get so involved that she would have to "go commercial with more workers helping". She lectures, and she creates Yuletide scenes in shops, hotels, and museums; some dolls are sold in an exclusive gift shop. The individuality and the delightful personalities of her Nisser are expressions of her own creativity.

In October 1989 we found the Christmas departments of Copenhagen's large department stores, *Illums*, and *Magazin du Nord*, totally overrun with Nisser! They appeared on gift wrapping, greeting cards, and paper napkins, and as tree ornaments. Children's coloring books and story books were abundant. A current book, *Nisserne pa Grønland* [54], takes them out of their hiding places and on a merry adventure to Greenland to find the lost Nisser who must carry the children's letters to the Julemand.

Nisser dolls of all sizes were everywhere in the stores. The largest, about 30 in (76. cm), seemed very real, standing in corners or leaning down from the highest shelves. Lack of luggage space dictated that we buy only the 11 in (27.9 cm) version. This happy creature's head, hands, boots, and socks are molded of soft vinyl. He wears a Nordic patterned sweater instead of the usual plain red garment.

A pair of 10 in (25.4cm) cloth Nisser and small knitted dolls were made for us by Mrs. Ester Pedersen of Kastrup. This 73 year young artisan had learned doll making in a class after convalescing from a broken back. She has made cloth Nisser with tricot faces over plastic molds, and small knitted Nisser. The pair is shown with the 11 in vinyl Nisse (Ill.110). The Julemand she made with bisque porcelain head and limbs is dressed in a modish zippered red jumpsuit with outsized pockets on the sleeves. The work of this artisan will be summarized in Chapter 12.

The last Nisse found in October 1989 is a mere bit of whimsy, a little plastic ball for a head on an opened beech nut husk wearing only a red pointed cap. Suspended by a string at its tip, he is to hang on the Christmas tree (Ill.111). A Danish grandmother in Colorado makes Nisser of peanuts and pipe cleaners; she arranges them on a small birch log.

The Nisser are a special entity in Danish folklore, whether dolls, figurines, candle sticks, or decorative articles. The collector may choose an antique Nisse, a reproduction of an antique, or a contemporary one — a DAM troll, a doll or a character figure — to give an added dimension, uniquely Scandinavian, to the Christmas decor and to the Christmas spirit.

Ill. 110

(110) Modern vinyl Nisse bought in Copenhagen in Oct.1989. with pair of cloth Nisser made by Mrs. Pedersen.

(111) Whimsical tree ornament Nisse with nut shell body, bought at Israels Plads flea market, Oct. 1989, with Nisse violinist of metal.

Ill. 111

Chapter 6

Cloth Dolls

A cherished rag doll is frequently the heart and soul of a child's play experience. The Raggedy Ann and Andy dolls beloved by generations of American children are typical examples. Cloth dolls have become only one of many artistic expressions of doll artists in Denmark and other countries today. Two hundred years ago, however, the dolls Danish children had were probably home made cloth dolls. Their creation depended upon the talents of the women of the family and the scraps of fabric on hand.

An example of a very early doll in the Danish National Museum is described by Sigsgaard and Varnild [55]. This doll made of rags is dated ca 1740, and is remarkably well preserved. Less than 5 in (9 cm) tall, it is called a miniature work of art. The drawn face is still clear as is the color and print of the clothing. The blue and white dress of printed material has a snug bodice and a long full skirt. The full sleeves are gathered into tight cuffs. A bright floral patterned apron which looks like linen covers the skirt. The yellow kerchief over the shoulders ties under her chin. This small but significant doll represents the early home made dolls in existence one hundred years before dolls were manufactured in Denmark.

The cloth dolls were probably made in small (under 6 in) to medium sizes because of the cost and scarcity of fabric. Small scraps would have been carefully utilized. Those home made dolls never went out of style; children loved and wore them out.

Dolls made in the homes as early as 1900 have been described by older Danish folks. Those might have been simply cut and stuffed with wool, sawdust, or even rag scraps. Often a baglike shape would have the head and waist delineated only by tightly wrapped string. Limbs would have been tight sausage-like rolls of fabric. And scraps of knitted yarn unraveled to make curls, or dots of wool sewn on the head became hair. The face was painted or colored with ink, or features were stitched with a bit of colored thread. Finding such a doll today would be unlikely; they were cherished toys and were loved to pieces. However, one lucky patron is the owner of a rare Danish homemade reversible doll, about 8 1/2 in (21 cm), dated 1910. It was sold in May 1990 at the International *Legetojsauktion* [56].

During World War II illustrated patterns with instructions were available commercially. Four examples are the Dagny-Doll, a Dutch boy, a topsy-turvy with two heads, and a nurse with unraveled yarn hair and a white apron [57].

A 17 in (43.2 cm) cloth doll we found in an antique shop in 1976 (Ill.112) represents efforts of one Danish lady to help the war effort. She made dolls of whatever fabrics could be found and sent them to a radio station, from which they were raffled for relief funds. The doll we have named "Hope" is cotton stuffed and has wiring in the long arms and legs. Face and hands are brown tricot. Eyes, eyebrows, and mouth are painted; the nose is a tightly stuffed round of tricot. Black cotton yarn made the heavy wig with bangs. Her stocking cap might have been a child's patterned mitten, cut and gathered into a tassle. Her red sweater is a scrap of knitted cotton. Draped around her shoulders is a bit of aqua colored crocheted wool, fastened with three little buttons and a piece of gold colored chain. The gathered skirt is brightly striped ribbed cotton, over which is an apron of black and white bold printed pique. Suspenders of aqua knitted fabric are attached at the waist where the striped skirt and apron are gathered and sewn to the doll. Tubes made of crocheted white wool are her stockings; sewn on shoes are pink with red soles. Red and white striped knitted cotton was used to make the bloomer styled underpants. And a frivolous touch is the petticoat of a piece of gold colored lingerie tricot edged in matching lace. Those undergarments, too, are sewn to the doll's body at the waist. This very colorful and dear old doll has survived the war and four and a half decades since the German occupation of Denmark. She is a real historical treasure.

Four cloth dolls which are among contributions to the Danish Immigrant Museum in Iowa were made in Denmark for friends and relatives in the U.S. One wonderful old knitted doll sent in 1921 to grandchildren was named "Funny Denmark" (Ill.113). The sewn on red cap probably made him a Nisse; if dressed as those described in the previous chapter he could be called "Funny Nisse". The doll's body is an oval with arms and legs attached. The face has simply embroidered eyes, eyebrows, nose and mouth. This doll was donated to the museum by Muriel Kruse Connolly.

Three other dolls made in Denmark ca 1977 by Karen Jensen were sent to the children of Ellen Modeweg in Illinois. These are 12 3/4 in (32.5 cm) to 13 1/2 in (34.3 cm) tall, and might represent a bearded father, a mother and a pigtailed daughter. The heads are stuffed stockinette with sparkling embroidered eyes, eyelashes, eyebrows, and mouths. Noses are merely tiny bumps in the stuffed faces. The man's hair and beard are sewn on curls of raveled yarn. The mother and daughter wear tight fitting, lace trimmed caps over hair of snipped knitted fabric. More braided knitted fabric forms pigtails five inches long, which are stitched onto the center parted hair of the girl. The man wears sewn on boots with trousers

Ill. 112

Ill. 113

Ill. 114

Ill. 115

(112) Cloth doll Hope made of scraps, for W.W.II relief funds.
(113) Funny Denmark, a Nisse, 1921. Photographed with permission June Sampson, Dir. Danish Immigrant Museum.
(114-115) Three cloth dolls, 1977, presented to Danish Immigrant Museum. Courtesy of June Sampson. Don Luckey photo.

Ill. 116

Ill. 117

(116) Mrs. Pedersen's dolls, 1989.
(117) Miss Blue Pedersen, 1989.

tucked into them. The mother's skirt and lace apron are longer than those worn by the girl doll. Lacy patterned knitted fabric, probably hose or tights, was used for the stockinged legs of the two. The slippers with shaped soles are sewn on at the ankles. The parent dolls wear modish buttoned vests over woven fabric shirts. Her forearms are rolls of stuffed knitted fabric; his are covered with heavy mittens. These three dolls might be called relatives of our Hope doll, having been created of an assortment of available fabrics, lace, and knits. The trio (Ill.114, 115) was photographed with permission of June Sampson at the Danish Immigrant Museum headquarters in April 1990. They appeared to be as expressive and alert as when made over 15 years ago.

In a 1983 series titled "Wee Immigrants" [58], two dolls marked "MADE IN DENMARK" closely resemble English made cloth dolls. A 9 in (22.9 cm) baker has a felt face; his wide grin shows painted teeth. His hands are felt, and his feet are carved wood. The doll wears a baker's cap and coat of white over black and white pants which are stuffed to form his lower limbs. The second doll, a 10 in (25.4 cm) fisherman, has a molded face of reddish felt, a large nose, and a pipe in his closed mouth. He sports a grey beard and heavy sideburns. This doll wears a yellow slicker and sou'wester, and high boots. The author, Margaret Groninger, proposed that the faces might have been created in the English Nora Wellings molds because of the similarities of the wide grin with teeth, and the big nose like those of the Chad Valley Seven Dwarfs. Her second suggestion is that these men dolls were part of a series representing specific Danish occupations. No similar dolls of cloth have been found, although contemporary jumping jacks of wood representing men's various occupations in Denmark will be presented in Chapter 11.

After World War II Miss Kirstine Larsen, born in 1891, made very personalized dolls with faces molded of many glued layers of fabric [59]. Were her dolls the inspiration for the recently fashioned facial molds used by many contemporary doll artisans? The very light weight plastic mold of the face and ears is designed to be covered with a stretch fabric which is shaped and stuffed to create the doll's head.

Aase Breivik's Danish and Norwegian folk dolls and those made by her pupils have beautiful faces made over such molds, and cloth bodies and limbs [60]. Her pair of Salling folk dolls appeared in Chapter 3. Dolls sold at the Israels Plads flea market in 1989 (Ill.116) were similarly made by Mrs. Ester Pedersen. Her 19 in (48.3 cm) doll bought there (Ill.117) was made entirely of light tan knitted fabric, softly stuffed. The paddle shaped hands were stitched to create fingers and thumbs; the soft feet wear boots. This lovely doll's artfully tinted face and features are emphasized by thick, lustrous medium brown curls, a purchased synthetic wig. Mrs. Pedersen had knitted the doll's light blue ensemble of tasseled stocking cap, turtle neck pullover, and tights, and navy blue boots. We have named the doll "Miss Blue Pedersen". Shown with two tiny dolls, simply knitted suits over plastic balls, are the pair of 10 in (25.4 cm) Nisser (Ill.118) which she made later utilizing facial molds. In a more complex body construction the limbs are attached at the sockets on cardboard discs, providing more articulation and easier posing. Miss Blue's limbs are merely sewn to the shoulders and the lower torso, with the resulting floppiness a part of her charm.

A little 5 in (12.7 cm) doll photographed with Miss Blue (Ill.119) was bought in a toy shop near Charlottenlund in 1978. Constructed simply of four pieces, the red and white striped body and unbleached muslin head, it becomes a baby's clutchable toy. Tiny eyes and mouth are stenciled, and wisps of red cotton yarn make a bit of a wig. The little bumps of arms and feet suggested we name it the "Danish Dumpling Doll".

In 1978 we also discovered printed kits to be made into dolls and other toys, copyrights of which are guarded vigorously. After making up commercially printed kits into toys for her children, Mrs. Jette Viby of Holte designed soft animal and doll kits, games, and wall hangings. They are marketed in Copenhagen shops, where toys made from the kits are displayed. Probably as popular as the soft dolls are the dachshund and her puppies, and the sow and piglets. On the mother animals the balls of snap fasteners become the udders, and the young have the sockets sewn on as mouths to suckle. We had made the dog family (Ill.120) and the doll Millie with her own matching wee doll and teddy (Ill.121). Another is the Turn-About Doll (Ill.122 a,b), a wide awake doll with its two small dollies; all reverse to a sleeping trio dressed in another color. Mr. Erling Andersen, now the manufacturer and sole agency of VIBY DESIGN at his Toftlund, Jutland factory, has sent us the completed doll, no.4001. Other popular toys of Mrs.Viby's designs include sets of finger puppets, a hobby horse on a stick, the Three Bears, and game board books such as *The Tinder Box*, complete with the characters as dolls and dogs to act out the story. Jette Viby and her accomplishments are included with other doll artists in Chapter 12.

(118) Mrs. Pedersen's Nisser pair, 1989.
(119) Danish Dumpling 1978 doll w/Miss Blue.
(120) Jette Vibe's dog with puppies. © Jette Vibe.

Ill. 118

Ill. 119

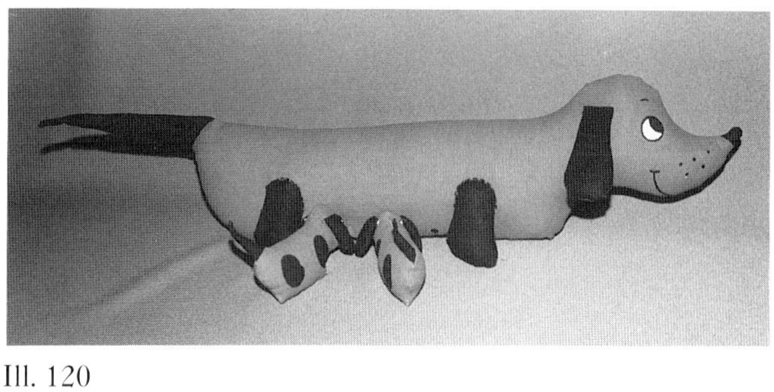

Ill. 120

Ill. 121

The last cloth dolls shown (Ill.123 - 125) are some of the needle sculpted creations by Pamela Nedergaard, who had made Nisser and folk dolls described earlier. These other delightful caricatures represent older peasant folk, appealing in quaint bonnets, spectacles, shaggy wigs, and moustaches on the men. The progression of styles and materials of Pamela's dolls, reflecting her development as a very talented doll artist, and the work of the other artists of cloth dolls will be summarized in Ch.12.

The charm and popularity of cloth dolls in Denmark today are very apparent. Doll kits, folk costumed dolls, Nisser, and caricature dolls are available in fine stores and boutiques as well as at the Israels Plads flea market, an outlet for licensed dealers. Increasing numbers of collectible dolls are created by artists in their home workshops and in classes they teach others.

Ill. 122 a

Ill. 122 b

(121) Millie. © Jette Vibe
(122 a,b) Turn-About doll, Jette Vibe design, © Bendy Color.

Ill. 123

Ill. 124

Ill. 125

(123-125) Soft sculpture dolls, courtesy of Pamela Nedergaard.

Chapter 7

Composition Dolls

Composition is the term used for a substance similar to the earlier papier mâché (Ch.11) but with sawdust, plaster or other ingredients added to the mixture of paper pulp and glue or resin. The resulting doll parts are usually thicker and stronger than those of papier mâché . Therefore the composition dolls might be expected to be longer lived.

Finding and identifying Danish made dolls of composition was the most difficult challenge within this research. The first and oldest is the 1926 birthday doll of Gunhild Nielsen (Ill.126). She is not sure whether it was Danish or German; it resembles the German bisque heads of the dolly face dolls popular for many years. The doll is not marked; might it have had a paper label giving its origin? By 1891 all exported dolls were required to bear the name of the country of origin. Most German composition dolls were marked or bore paper labels, but unmarked German dolls not made for export might have been brought into Denmark from Schlesweg, suggested this doll's owner. All the old composition dolls we have found in Germany are clearly marked on the heads and sometimes on the bodies (Ill.127). The origin of Gunhild's lovely doll, a resident of Denmark for over sixty years, remains an unknown; was it Danish or German made?

Mrs. Doris Poulsen, Director of the Randers Doll and Toy Museum, sent a photograph of a boy and girl pair displayed there (Ill.128). These attractive black dolls, almost 17 in (45 cm) tall, were made by the *Keramishes Werk* (Ceramic Works) ca 1932. They are the earliest composition dolls which a Danish expert has positively dated and identified. The heads are composition; the bodies are papier mâché . Flirty eyes, open-closed mouths showing tongues and teeth, and thick black wigs distinguish this very black pair of rare dolls she named "Sarah and Tom".

Another boy and girl pair (Ill.129) was photographed at the Copenhagen home of doll collectors I. and P. Pedersen. The dolls are unmarked. The concensus of the sellers and the collectors was that they were made in the 1930s, and were Danish, not German. They resemble Kammer and Reinhardt's popular Peter and Marie, and wear their original *lederhosen* (leather pants) and dirndl (Ill.130). The dolls are 12 3/4 in (32.5 cm) tall. The heads are composition; bodies and limbs are stuffed cloth.

Ill. 126

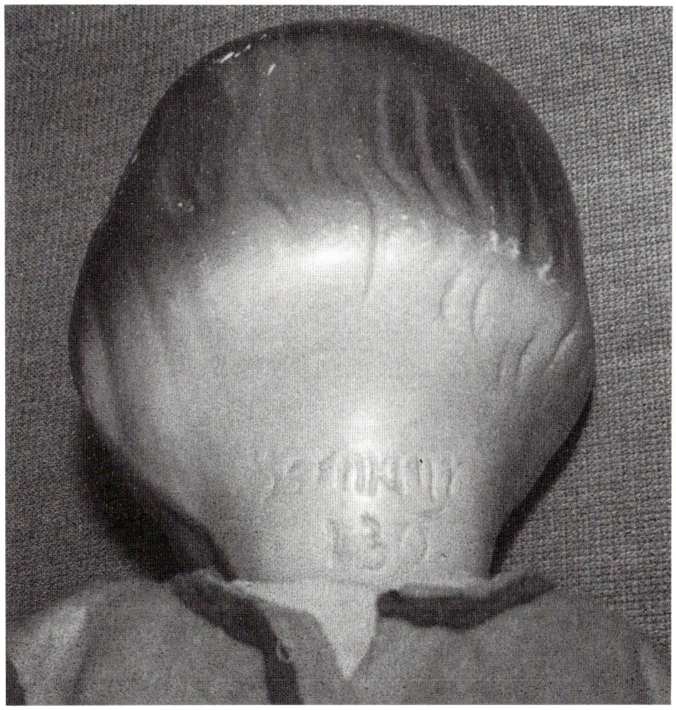

Ill. 127

Ill. 128

Ill. 129

(126) Composition doll of 1926. Courtesy of Gunhild Nielsen.
(127) German composition doll showing marks.

(128) Composition dolls, Sara and Tom. Courtesy of Randers Doll and Toy Museum.

(129) Composition dolls resembling German dolls by Kammer and Reinhardt. Courtesy of I. and P. Pedersen, Copenhagen.

Ill. 130

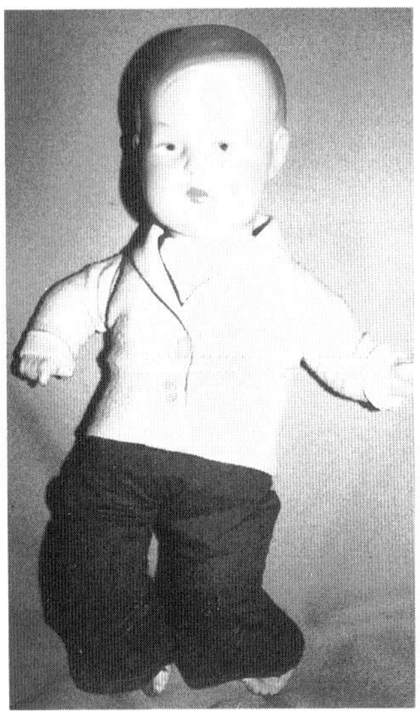

Ill. 131

Ill. 132

130) Details of dolls in Ill.129.

(131) Toddler boy doll in original costume. Courtesy of I. and P. Pedersen.

(132) All composition girl doll in original dress. Courtesy of I. and P. Pedersen.

Eyes and lashes are painted; the mouths are almost pouty. The boy's molded hair is painted reddish-blonde. The girl's mohair wig of the same color is arranged in a *frikadella* style, named for a popular coiled Danish breakfast bun [61]. A British doll authority, Maree Tarnowska, called this hair style, braids coiled over the ears, ram's horns on bisque dolls made by Heubach and by Kammer and Reinhardt [62].

Two more unmarked and undated dolls were shown to us by the Pedersens. The 12 1/2 in (31.8 cm) bent leg toddler boy (Ill.131) has composition head, arms, hands, and feet on a cloth body. The fingers and toes are outlined in red. His features are similar in color to the boy's in Ill.129. His original costume is of felt, a light grey jacket with tan trim and faded navy trousers. The rare all composition girl doll, 10 in (25.4 cm) tall, has clearly delineated facial features, blue eyes with white highlights, dark red-blonde eyelashes, red tear duct dots, nostrils and full lips. There are red tracings on her fingers and toes. Her molded hair is painted reddish-blonde. This doll wears her original polka dotted dress (Ill.132)

A doll almost 17 in (43 cm) tall is a single entry in the May 1990 international auction catalogue which might be classified as composition. It is identified as "probably 'Sorring' Denmark ca 1940". The description is "Clay-doll, boy with painted clay-head, closed mouth and painted blue eyes... Clay-body...". In Stuttgart, Germany a doll of clay-like composition called *Keramic-bisquit* (ceramic-bisque) had identified our Armand Marseille baby 2549. Our baby doll is classified a composition doll in an American price guide; no category such as ceramic-bisque seems to be used in American classifications. The boy doll is articulated at the shoulders and hips; his molding resembles that of a smaller metal boy described in Ch.11. The Sorring boy wears a two piece sailor suit with short pants buttoned to the shirt and more buttons on the sides of the trousers, short socks, and leather shoes.

Mr. Finn Risom, a very knowledgable collector in Aalborg, sent me the following detailed information [63] "about the (doll) industry in Denmark around World War II. After the war, up to about 1954, there were heavy restrictions on imports, which prolonged the situation (of) a protected market, with shortage(s) of almost all goods. The dolls produced at that time were often copied from imported, pre-war dolls. That means that one can't identify the doll's age from (its) hair style, etc. Of course most of the copying was not legal, which is perhaps why so many unmarked dolls were made. Because of the small scale of the production, (there were) no catalogues. Production stopped after (a) few years, and it is very difficult to identify (the dolls) found today from that period."

The following groups of dolls and doll heads are documented and shown, with Mr. Risom's kind permission. The three-faced doll (Ill.133 - 135) is considered an American design. Doesn't the doll look like

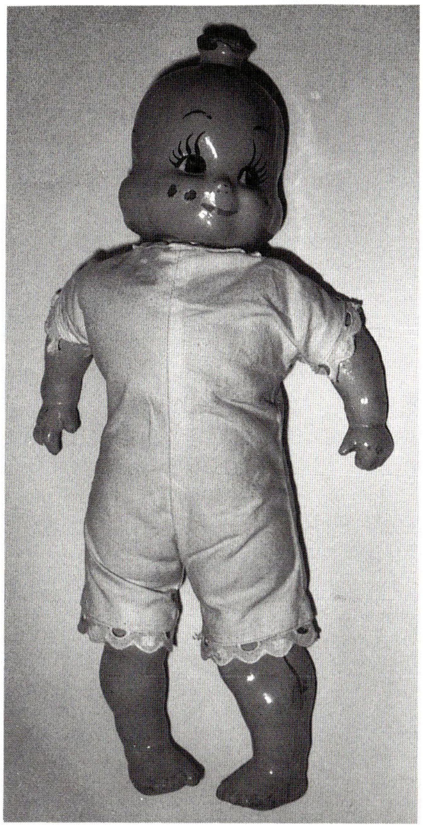

Ill. 133

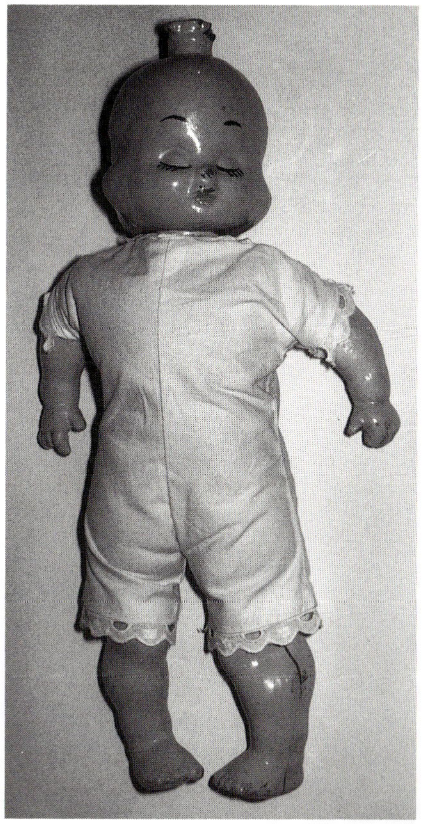

Ill. 134

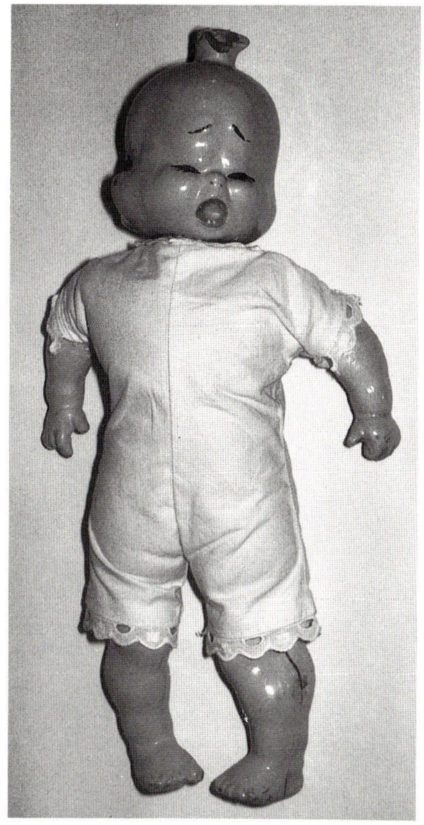

Ill. 135

Ill. 136

Ill. 137

(133-135) Three-faced doll, photos courtesy of Finn Risom, Aalborg.

(136, 137) JEMA doll heads of composition. Photos courtesy of Finn Risom.

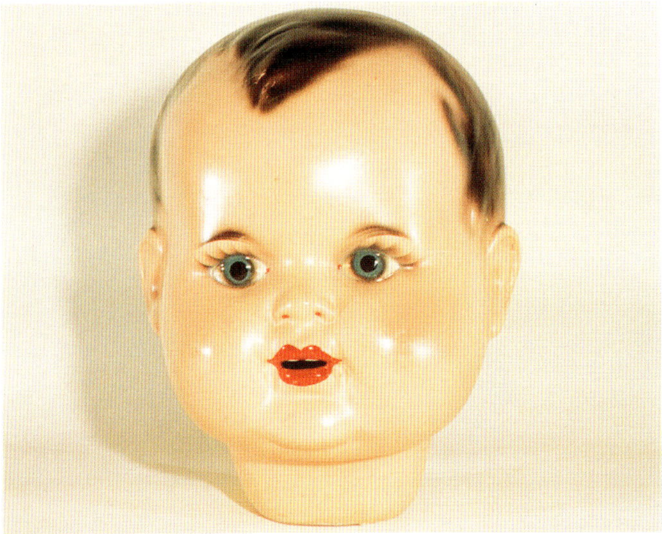

Ill. 138

Ill. 139

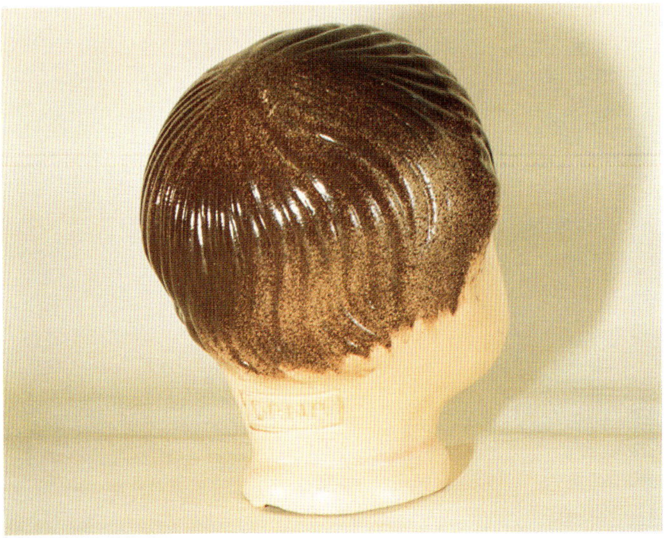

Ill. 140

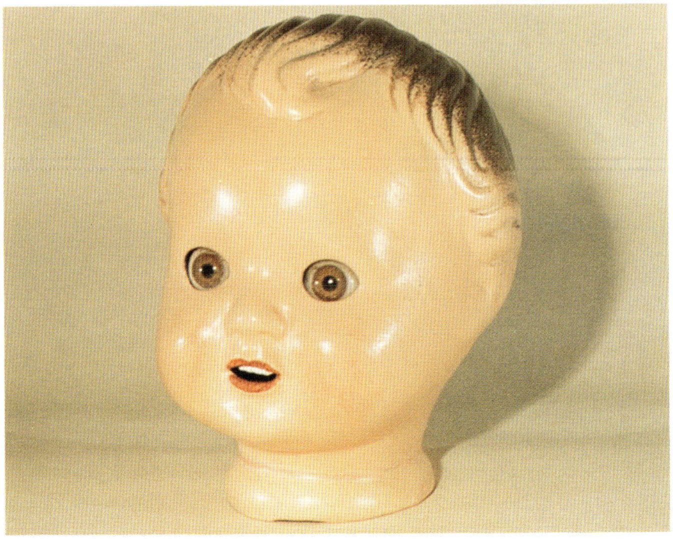

Ill. 141

(138) JEMA doll head, back and front, showing faint mark.
(139) Photos courtesy of Finn Risom.
(140) Back of another head showing faint DANA mark. Photo courtesy of Finn Risom.
(141) Front of DANA head. Photo courtesy of Finn Risom.

Elsie Gilbert's Trudy, made in New York City in 1946? Mr. Risom suggests this is the so-called Trio-doll in a 1947 Danish wholesale list: "Three different expressions, crying - sleeping - laughing, dressed in gown, bonnet and shoes".

Mr. Risom found four doll heads in a country shop; a doll shop owner of that time identified them as JEMA dolls. She had identified the maker as *J. E. MOLLER ANDERSEN*. The four doll heads (Ill.136, 137), unmarked, were further identified by comparing them with marked JEMA doll heads. All four are similar in facial shapes and especially strong colors of the painted features. The face of the all composition doll in Ill.132 somewhat resembles these JEMA dolls' looks. The fifth doll head he found later is plainly marked "JEMA" on the neck (Ill.138, 139). A sixth doll head is marked "DANA", scarcely visible in the photo (Ill.140); it has lighter coloring and sleep eyes of brown with no visible eye lashes (Ill.141). Mr. Risom has identified it as a product of the Dana firm at Vibevej 6, Copenhagen, from an advertisement dated 1948. He has seen two more Dana dolls and regretted not having photographs to send.

Mr. Finn Risom has made a very valuable and much appreciated contribution to the little known history of Danish composition dolls.

The last doll to be included here, found in October 1989, is a look-alike of the Kammer and Reinhardt 100, frequently called the Kaiser Baby. This baby doll has a composition head and arms on a cloth body (Ill.142), This 11 in (27.9 cm) bent leg toddler doll was undressed but is very clean, perhaps never played with. The resemblance to the well known German character baby might be explained by Mr. Risom's history cited above. It has rare metal legs which will be discussed in Ch.11.

The known dates of Danish composition dolls are from the early 1930s into the post-war period, beyond the mid-1950s. This time period parallels the production of Danish celluloids, from Nisser dated ca 1930 through the thriving 1950s industry built up by Jens Kofod. The beautiful celluloid dolls and doll heads are presented in the following chapter.

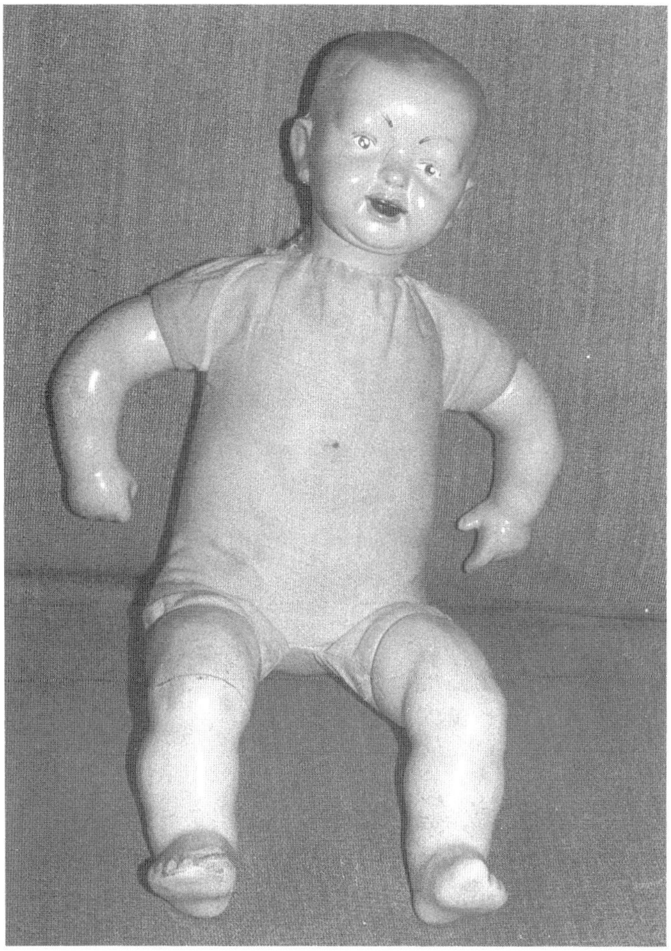

Ill. 142

(142) Doll with composition head and arms; a Kaiser baby look-alike.

Chapter 8

Danish Celluloid Dolls

Kader Dolls

The *Rheinische Gummi-und Celluloid-Fabrik* at Mannheim, now the *Schildkröt* Company, was the leading German manufacturer of celluloid dolls until "total destruction of the company, producing munitions, not toys, during both world wars" [64]. Those circumstances catalyzed celluloid doll manufacturing in Denmark in the 1940s. More than two prosperous decades followed.

The earliest all celluloid Danish dolls, previously unknown to us, were photographed and documented at the Copenhagen home of I. and P. Pedersen in October 1989. The marks on the back of the heads show the letters "OK" superimposed upon a slightly tilted globe. Beneath the globe is the word "KADER" (Ill.143). Who was Kader? Individual marks with letters and numbers will be cited with descriptions of each of four beautifully designed dolls.

A 20 1/2 in (52.1 cm) boy doll is marked just below the globe and name with "B3520" and "BH", and on the back with "BH". This doll has light brown painted hair with one lock waved in the middle of the forehead, deep blue sleep eyes, and an open-closed mouth with two upper teeth (Ill.144). He wears a long tailored shirt over trousers fitted at the ankles and woolen booties (Ill.145).

A pair of toddler babies, 10 1/2 in (26.7 cm) and 11 3/4 in (29.9 cm), shows the globe trademarks. These are numbered "B3510" and "B3512 1/2" respectively. The celluloid is darker than that of the other Kader dolls. The hair style of the boy, the taller doll, is the same color and style of the first doll in this series. He sucks a pacifier (Ill.146, 147). The girl's frontal lock is curlier than the boy's, but it is the same light brown color (Ill.148, 149). Both dolls have light blue, almost turquoise colored, sleep eyes. The toddler girl wears a crocheted dress of gold and red-orange; the toddler boy wears blue knitted short pants and a red shirt striped with blue and white.

Another all celluloid doll, a baby, 19 in (48.3 cm), is very pale in color and extremely light in weight. She, like the first doll, is marked "B3520" beneath the globe. This doll has more marks, but they are too indistinct to further identify these Kader dolls: "Made in - - - ." This beautiful baby doll has dark blue eyes which are remarkably well balanced and seem to always look directly at one. The hair is waved like that of the two boy dolls described (Ill.150). She is dressed in a ruffled bonnet, a

Ill. 144

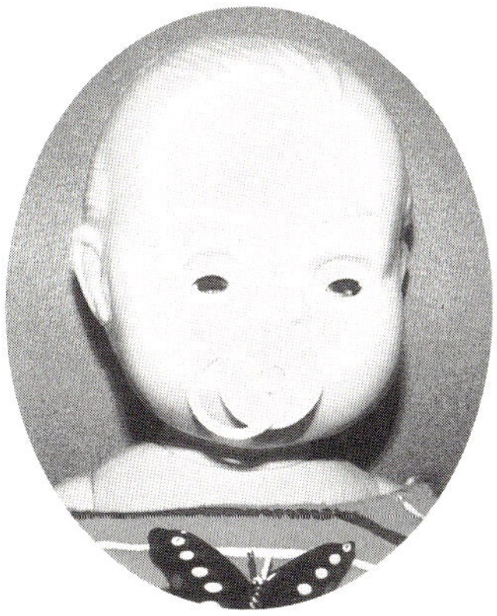

Ill. 146

Ill. 148

Ill. 150

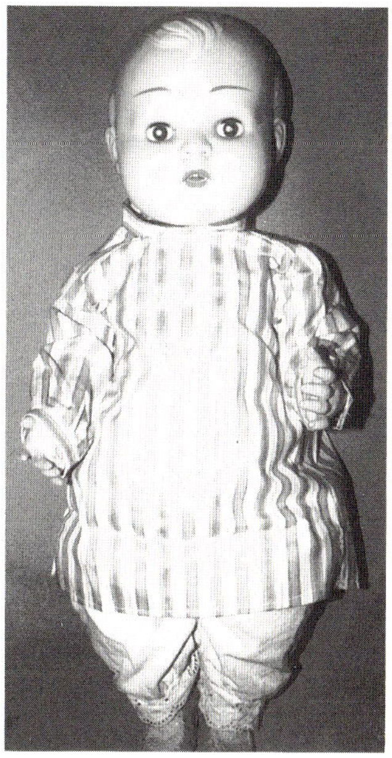

Ill. 145

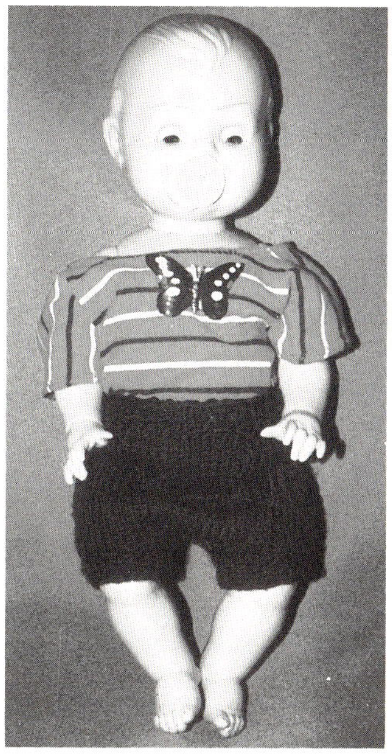

Ill. 147

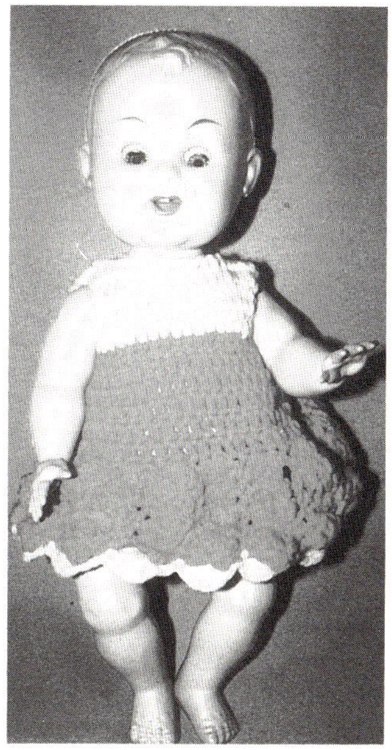

Ill. 149

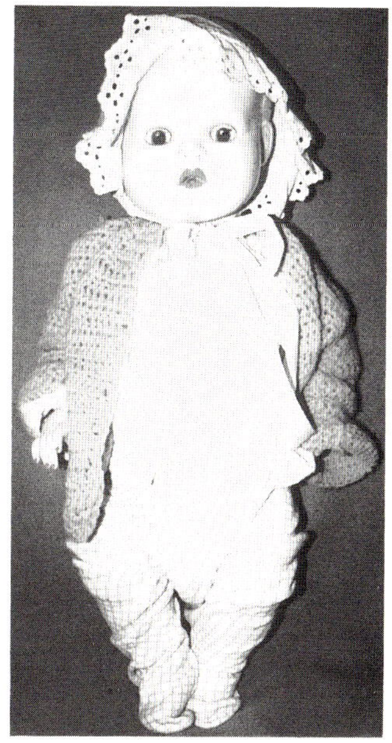

Ill. 151

Ill. 143

(143) Trademark showing globe and KADER. Sketch by E. van Baren.

(144) Close-up of large boy doll marked KADER. Dolls in Ill.144-151 are all in I. & P. Pedersen collection.

(145) Full length, boy doll.

(146) Close-up of toddler boy with pacifier.

(147) Full length of boy with pacifier.

(148) Close-up of toddler girl.

(149) Full length, toddler girl.

(150) Close-up of baby doll.

(151) Full length, baby doll.

Ill. 152 a

Ill. 152 b

Ill. 153 a

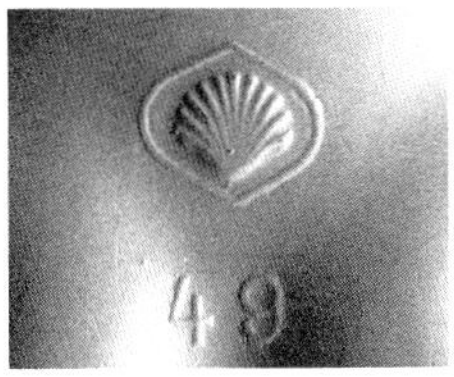

Ill. 153 b

Ill. 154

Ill. 155

(152-155) Marks of Køge dolls, 1940 - 1960. Sketches by P. Luckey.

light green sweater over a white blouse, and red checked tights (Ill.151).

Similarities are readily seen in these Kader dolls. The hair of three of them is styled with a wave in the center of the forehead. The painted hair is dark blonde or light brown, and the one-stroke eyebrows are medium brown. An unusual feature I have not seen in other dolls is the detail in the sleep eyes. Dots rim the iris of each eye, more distinct in the darker blue eyes. All four dolls have toddler bodies with articulation at the shoulders and hips. The hands of the dolls are well detailed with individually molded fingers.

There may be clues to identification of the dolls within the marks on each of them. The closeness of the last two numbers to the doll's height may indicate size. The first and the fourth dolls are numbered "B35**20**"; the heights are 20 1/2 inches and 19 inches. Similarly, the second and third dolls are numbered "B35**10**" and "B35**12 1/2**"; corresponding heights are 10 1/2 and 11 3/4 inches. The slight variation beteen the numbers in the marks and in the dolls' heights is from 1/2 to 1 inch. Measuring the toddler dolls, with legs in the bent position for sitting, could have been inaccurate, and could account for these slight differences. Denmark formerly used the term *tomme* (plural: *tommer*), which approximated the inch. The measurements inch and foot are used in Denmark today. If these five digits might include the size code the possibility of German manufacture is eliminated, because a German size number is expressed in centimeters. A second clue within the marks might be a date. According to information given the collectors by the sellers, these dolls were made ca 1938. Might the number 35 within the doll's mark suggest the earlier year 1935? We have no clue about the letters "B" in the code numbers and "BH". Could those be initials of the designer or manufacturer? Nothing is known about the word Kader on the dolls. In the town of Køge doll making flourished during the next two decades. Were these dolls made there? The history of the Køge doll industry, 1940 - 1960+, will cite several companies making celluloid dolls which were ultimately absorbed by J. Kofod.

Køge Dolls

An article prepared for the archives of Køge, a coastal city south of Copenhagen, has provided answers to more unknowns within our Danish celluloid doll research [65]. This Køge doll history identifies marks and dates on dolls we had found in antique shops, toy shows, and the Israels Plads flea market.

On land which he bought in 1940 in Køge, Johan Olsen Kofod, a delivery trucker also known as Jens Kofod, began a papier mâché doll factory and made dolls for the next four years. The production of papier mâché dolls, wooden dolls and toys, and doll prams increased rapidly after 1944 when he acquired the Prefa factory at Glostrup, near Copenhagen. Expanding further in 1946, he added papier mâché masks to the growing doll manufacturing. In that year he began making ceramic dolls and piggy and dog banks. Celluloid doll production was begun the following year, 1947, when German molds were bought in Norway, and others were bought from the Schildkröt factory. The celluloid doll business became very successful in just one year, but new mold designs were desired. Making the molds in the factory proved difficult; that was turned over to the sculptor Lomsen. The new molds were cast in bronze by Harry Andersen of Copenhagen.

The resulting beautiful dolls brought a director from the Schildkröt firm to Køge in 1948. He said that they were the most beautiful celluloid doll faces he had ever seen, very characteristically Danish, not copied or borrowed from any other factory. By 1950 molds were being made at a cost of 7,000 to 15,000 Danish Kroner each, which was then equal to $1,015.00 to $2,175.00 [66]. Celluloid was imported from Italy, and cryolite (limestone, used in the molding process) was shipped from Greenland. Since postwar production costs were very high, more dolls might have been sold in Sweden than in Denmark.

Through 1954 a triangle mark was used in the Køge factory. This mark is readily identified, with the city's name within the base of the triangle and above it the initials, J.K., of Johan (Jens) Kofod (Ill.152). The Swifty Celluloid Co. used a shell mark (called mussel in Denmark) within an elliptical frame (Ill.153) from 1952 to 1954. This firm was bought by Mr. Kofod in 1955. In that same year his firm bought the Danish Celluloid Industry, which used its initials, D.C.I., within a heart (Ill. 154). Mr. Kofod chose to retain the mussel mark but without its frame (Ill.155).

Between 1955 and 1957 many celluloid, plastic, and rubber dolls and bathtub toys were produced. As many as 4,000 rubber balls were made daily in 1957, and the business continued to prosper. The Schildkröt firm eagerly acquired shares in Kofod's firm. Hans Stab at age 28 became the very successful German Director.

In 1959 a new department was created to produce soft vinyl dolls, and a new material, *celludukke*, was used for the baby dolls. Other dolls were still being manufactured of celluloid and of rubber. Hans Stab, still in charge of the Schildkröt production at Køge, opened a sales department in Copenhagen under Otto Nielsen. This became the *A/S Tøj og Hobby Scandia* (R Toy and Hobby Scandia). By 1960 it was a thriving business, with Schildkröt sharing in the production for the Danish market.

J. O. Kofod died in 1967. The doll and toy factory was sold to *Rheinische Gummi und Celluloidfabrik* in the early 1960s; the Danish newspaper *Dagbladet* reported its sale

Ill. 156

Ill. 157

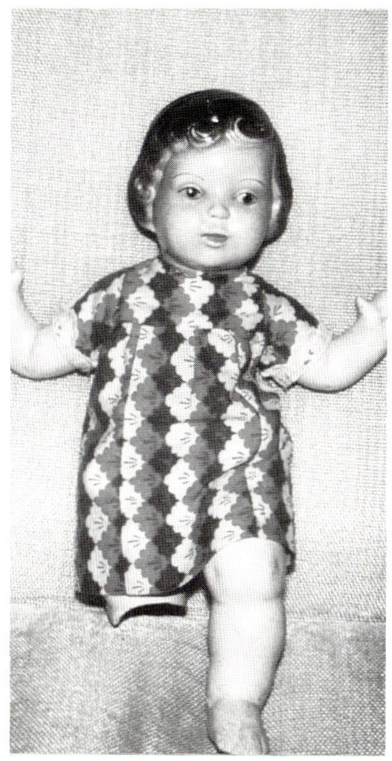

Ill. 158

(156) Celluloid doll on leather body, close-up. I. & P. Pedersen collection.

(157) Full length of celluloid lady showing elegant dress.

(158) Early celluloid head on cloth body; limbs of ceramic-like substance seen at broken toes.

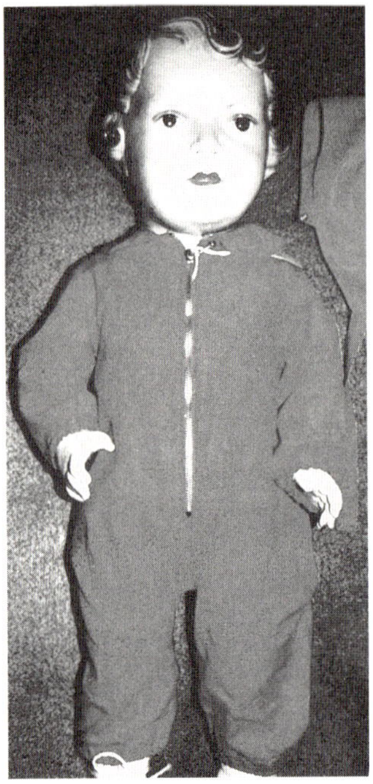

Ill. 159

Ill. 160

Ill. 161

(159) Largest doll, 23 in., in original red suit with cap.

(160) Close-up of largest celluloid boy.

(161) All celluloid girl, with enclosed Mussel mark, wearing hand knitted blouse, coat and leggings of pink wool.

back to a Danish toy dealer, Mr. Poul Andersen, of Soborg near Copenhagen, in 1963 [67]. We have found no Danish dolls which have the turtle mark combined with a Danish Køge mark. However, a small vinyl doll with the turtle mark of Schildkröt appeared in a Danish advertisement in 1970 which included the name "Kofod", to be detailed in Chapter 11.

We are estimating that the earliest Køge doll documented here is the 19 in (48.3cm) fashionably dressed doll, with a celluloid head on a tall, slim leather body. Her leather hands and wrists are on cloth arms; her feet are leather. Uneven color in her pale face suggests fading. Her reddish blonde hair shows deeply sculpted waves and curls. This very stylish lady is dressed in a costume suggesting a walking suit, a white silk blouse, a jacket and long skirt of red and white checked rayon, a hat, a long scarf, and a parasol of ecru lace (Ill.156, 157). The doll is marked with the triangle.

Our earliest Køge doll, a 14 in (35.6 cm) girl, was found in May 1989 at the Israels Plads flea market, in a box of numerous celluloid doll parts and broken bodies. Her celluloid head shows the same rather serious expression and molding of the fashionable doll in the Pedersen collection. Both have well defined brown eyebrows and eyeliner, blue eyes, and red tear duct dots and nostrils. Our girl doll's shoulder plate head and crudely formed hands, arms and one leg are on a white cotton body. Where the right leg and left toes are broken off, a type of plaster is visible. The arms are painted white, the legs pale flesh colored, in contrast to her tan face. Her wavy molded hair is shaded from medium to dark brown (Ill.158). The triangle marked celluloid heads of the two look-alikes and the material of the legs and feet of the second doll might date them as early as 1947.

The first and largest celluloid doll, 23 in (58.4 cm), we found in 1984. The socket neck boy doll wears his original clothes, a red suit and cap. Some facial paint had worn off his chin and his forehead curls revealing the yellow tone of the celluloid base. He has green eyes and faint brown eyebrows. The tear duct dots and nostrils of red are details of most of the "J.K.//KØGE" marked dolls we have bought (Ill.159, 160). Our second doll was the 19 in (48.3 cm) socket neck girl in pink knitted coat and leggings (Ill.161). Her mark is the enclosed mussel of the Swifty Company. And there was the challenge; the antique dealer told us the girl doll was also a Køge doll! Immediately evident is the variation in the brown hair styles, although the facial details of both dolls are similar. The Køge history we learned five years after we found these two dolls explains the Køge label identified by the sellers for both the triangle and mussel marks.

Our next early purchase was a pair, a 9 in (22.9 cm) celluloid doll and a 3 in (7.6 cm) shoulder head (Ill.162). The hair styles of these triangle marked heads are almost identical, but differences in facial painting are apparent in the close-up (Ill.163). This all celluloid doll has an

Ill. 162

Ill. 163

Ill. 169

Ill. 164

Ill. 165

Ill. 166

Ill. 167

Ill. 168

(162) All celluloid doll with shoulder head of similar look.
(163) Close-up of dolls, Ill.162, showing differences in coloring.
(164) Our most recently purchased celluloid doll heads.
(165) Blonde doll in knitted Norwegian costume, as sold over 30 years ago.
(166) Doll of Ill.165 in dress of old Danish hand woven fabric.
(167) Silver haired doll, collection of Pamela Nedergaard, Odense. (All P. Nedergaard's dolls photographed and documented with permission, at Odense, May 1989.)
(168) Doll with gold painted hair and rare sleep eyes. P. Nedergaard collection.
(169) Full view of Pamela's dolls, 167, 168, with metallic hair.

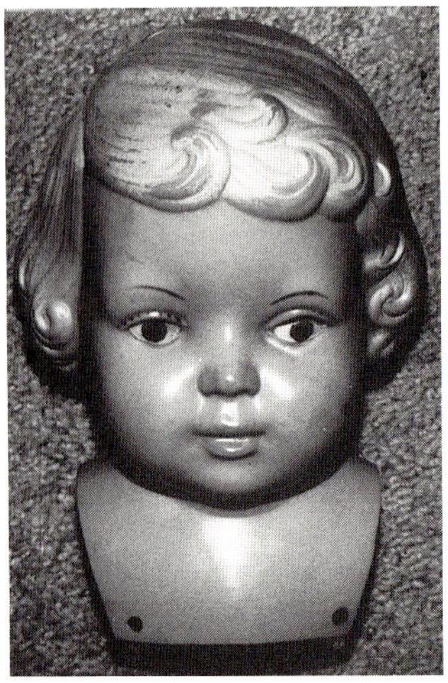

Ill. 170

Ill. 171

Ill. 172

Ill. 173

Ill. 174

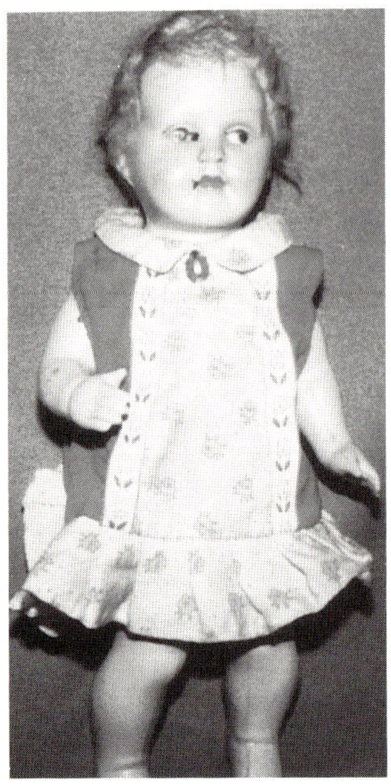

Ill. 175

Ill. 176

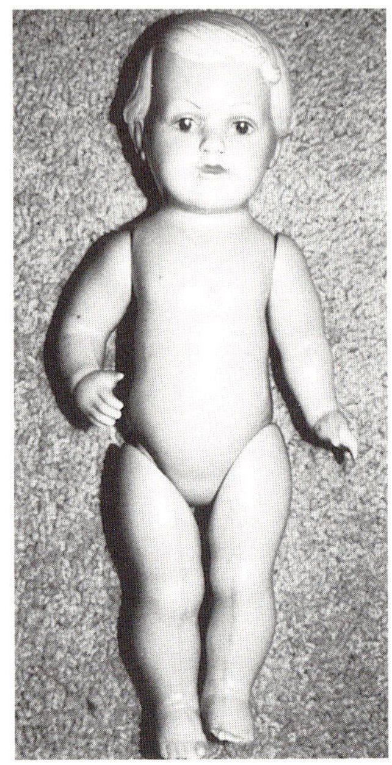

Ill. 177

(170) Doll head with flat white painted hair. P. Nedergaard collection.

(171) Doll with sleep eyes and pale gold-glossed yellowhair. Photographed courtesy of Kirsten Beck at her antique shop, *Boutique Belle Epoque*, Copenhagen, May 1989.

(172) Rare all celluloid doll with sleep eyes, socket neck, and wig. I. and P. Pedersen collection, Copenhagen.

(173) Full length of doll in Ill.172.

(174) Second wigged girl doll, close-up. I. and P. Petersen collection.

(175) Full length of doll in Ill.174.

(176) Doll with brown hair. P. Nedergaard collection.

(177) Undressed 9 in. pale yellow haired doll. P. Nedergaard.

extensive hand knitted wardrobe which includes a green and grey jacket and cap, a dark green coat, a blue and pink bonnet, a blue dress, a dark green vest, an outdoor set of blue jacket, mittens, and leggings, and a 2 piece blue and white dress. The little doll and her wardrobe must have been a treasure to its doll-mother.

Other doll heads bought in October 1989 (Ill.164) show similarities in the molded hair of different colors. These heads, all with the Køge triangle, have never been on doll bodies.

The 18 in (45.7 cm) socket neck blonde doll wearing a long knitted dress also has the triangular Køge mark (Ill.165). This was a childhood doll of Benedicte Nielsen, bought by her aunt at a church bazaar before 1960. The doll, not new, had been dressed for resale in a costume of Norway.

> "You may have her if you like her. I never played with her because she wasn't as pretty as my other doll with silver hair."

Happily we claimed the doll in May 1989, at Charlottenlund. Of course this doll is very special! We were delighted to find that the very old dress of hand loomed cotton bought years before gave back her Danish authenticity (Ill.166).

We had supposed that Benedicte was describing a white haired doll until we saw the beautiful celluloid dolls in Pamela Nedergaard's collection. Her 15 7/8 in. (40 cm) doll dressed in red, with lustrous dark brown eyes and silver hair, (Ill.167) is the most beautiful of all the Danish celluloids we have documented. Her 19 1/8 in. (48.6 cm) doll with blue sleep eyes has gold painted hair (Ill.168). Yes, both dolls have metallic hued hair, very unusual and very lovely! Both have the later mussel mark. They are shown together in full length (Ill.169). The taller doll wears the uniform of the Danish Girls' Brass Band, a short jacket and skirt, and fur cuffed marching boots, her original costume. Contrasted with these metallic hair colors of silver and gold is the flat white hair on another doll head (Ill.170).

Sleep eyes in Danish celluloid dolls might have been very rare. Another doll with very light brown sleep eyes and metallic glossed pale yellow hair was found and photographed in 1989 (Ill.171), courtesy of Kirsten Beck.

A celluloid doll with sleep eyes and a wig of long brown hair is a most unusual doll. This is the third and last doll we found with sleep eyes, one of only two with wigs, and the only wigged doll with a socket head (Ill.172.) Her blue sleep eyes, red tear duct dots, and the red tracings of the fingernails and toenails are marks of a finely made doll. The cost of finishing dolls with these details, common on many German dolls, might have been a factor in the scarcity of this model. Production costs of all celluloid dolls with socket neck construction would have been higher than of dolls with heads molded with the body. This mussel marked doll was a later model, possibly

Ill. 178

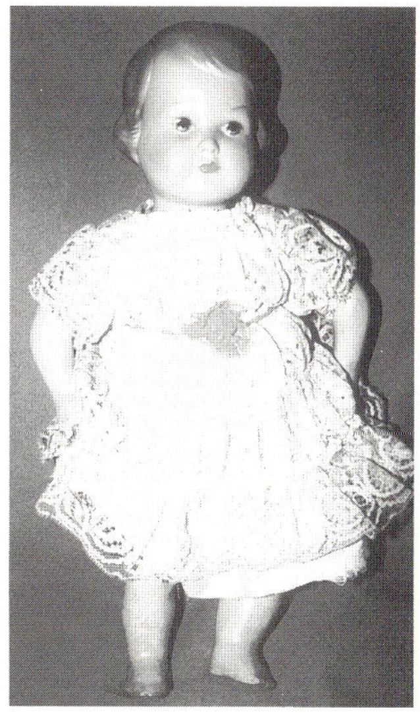

Ill. 179

Ill. 180

Ill. 182

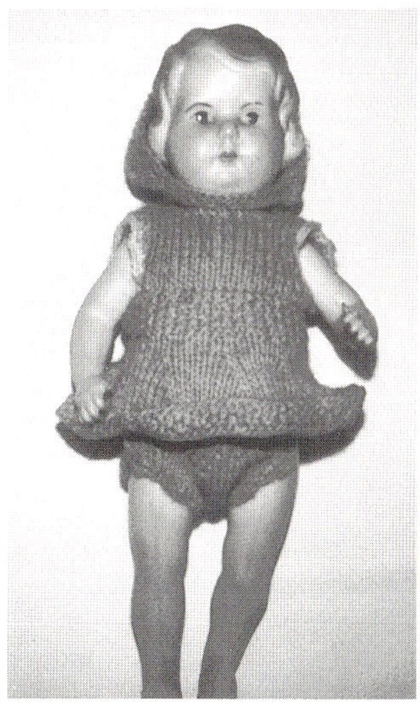

Ill. 181

Ill. 183

(178) Close-up, green eyed doll resembling Schildkröt dolls. I. and P. Pedersen collection.

(179) Full length doll in Ill.178, showing original dress.

(180) Close-up, doll resembling a Schilkrödt celluloid doll. I. and P. Petersen collection.

(181) Full length doll in 180, in hand knit cothes.

(182) Close-up of lighter colored twin to Ill.181. I. and P. Pedersen collection.

(183) Full length of Ill.182, showing hand knitted dress.

Ill. 184

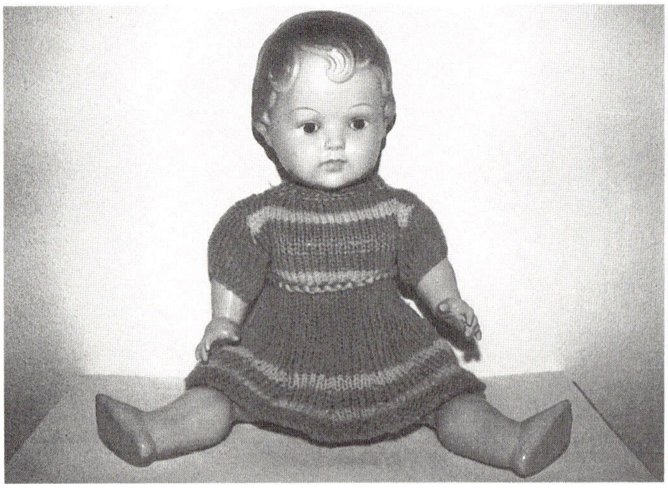

Ill. 185

Ill. 186

(184) Close-up, author's last all celluloid doll.
(185) Full length of Ill.184.
(186) Rare black baby doll by Kofod. I. and P. Pedersen collection, Copenhagen.

experimental, made by Mr. Kofod. The Pedersens told us that this beautiful girl doll wears her original dress (Ill.173).

A second wigged doll, 9 1/2 in (24.2 cm), triangle marked, was another rare doll in the Pedersen collection, an earlier all celluloid doll. The facial details had been well painted; unfortunately the left corners of her eye and her mouth had been punctured. The dark blonde wig may have been cut. In spite of abusive treatment, the doll still wears her original dress, red with red and white printed collar, front panel, and ruffle at the hem (Ill.174, 175).

Two more all celluloid dolls photographed at Pamela Nedergaard's home have the mussel mark. One 14 in. (35.6 cm) girl doll in a blue dress has very blue eyes and medium brown hair (Ill.176). A 9 in. (22.9 cm) undressed doll has bright blue eyes and light yellow blonde hair (Ill.177).

A 13 in (33 cm) green eyed girl doll has molded hair very similar to that of many Schildkröt celluloid dolls. We know she is not German; she has the triangle mark. The doll is wearing her original lace dress (Ill.178, 179). Two other all celluloid dolls which are unmarked also resemble Schildkröt dolls. The darker colored 9 1/4 in (23.6 cm) girl has vividly colored painted features, and wears a blue-grey knitted ensemble of a dress with attached hood, and panties (Ill.180, 181). The look-alike doll, 9 3/4 in (24.9 cm), is of lighter celluloid, and her features and hair show some fading. She, too, wears a hand knitted dress with bright stripes creating the full skirt and trim on the sleeves (Ill.182, 183). The full faces and double chins of these unmarked dolls have a German look, but if they were made by Schildkröt they would probably have had the turtle marks. All our Schildkröt celluloids are clearly embossed with the turtle. Were the look-alike dolls made from the earliest molds the company had acquired from Schildkröt? The Pedersons agreed that they are probably Kofod dolls.

The 13 in (33 cm) girl doll with dark brown wavy hair and very dark blue eyes is the last all celluloid doll we bought in Denmark. This doll has articulation only at the shoulders and hips, characteristic of most of the larger dolls and all the doll house dolls. The triangle marked doll is wearing a hand knitted dress of blue with rose and aqua stripes, the skirt attractively flared, and blue knitted panties. A crocheted belt combines several colors (Ill.184, 185). Obviously many dolls were dressed by the needle-workers of families or friends who skillfully and economically used up bits of leftover yarns.

A very exciting discovery was the black baby doll, 10 in (25.4 cm), the only black one we have found which was made by Kofod (Ill.186). He is dressed in machine knitted shirt and tights, and a gold crocheted vest closed with four buttons. This triangle marked doll is another treasure of the Pedersens.

Doll house dolls we have documented are 6 1/8 in. (15.5 cm) tall. Two of the little dolls have Dutch boy bobbed hair, and molded shoes and sox. Pamela's mussel

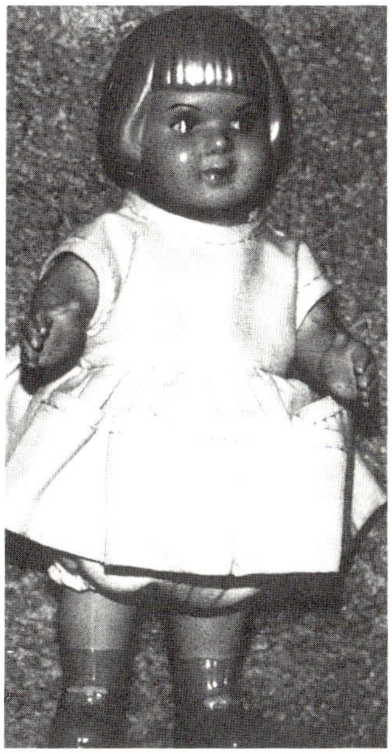

Ill. 187

Ill. 188

Ill. 189

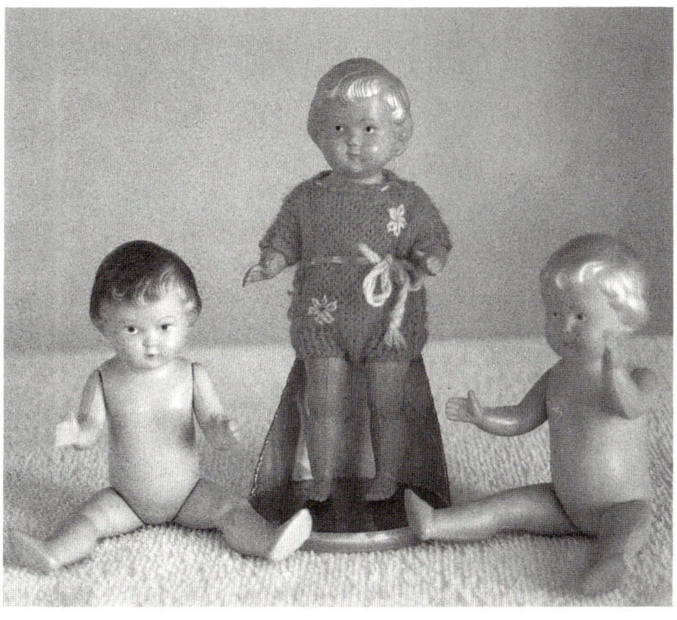

Ill. 190

Ill. 191

(187) Doll house doll, blonde Dutch bob hair style. P. Nedergaard collection.

(188) Brownette twin of Ill.187; a 1989 find by author.

(189) Two triangle marked doll house dolls.

(190) Three doll house dolls, look-alikes with different marks, different coloring.

(191) Pair of wavy haired doll house dolls with two marks; photographed courtesy of Kirsten Beck, 1989.

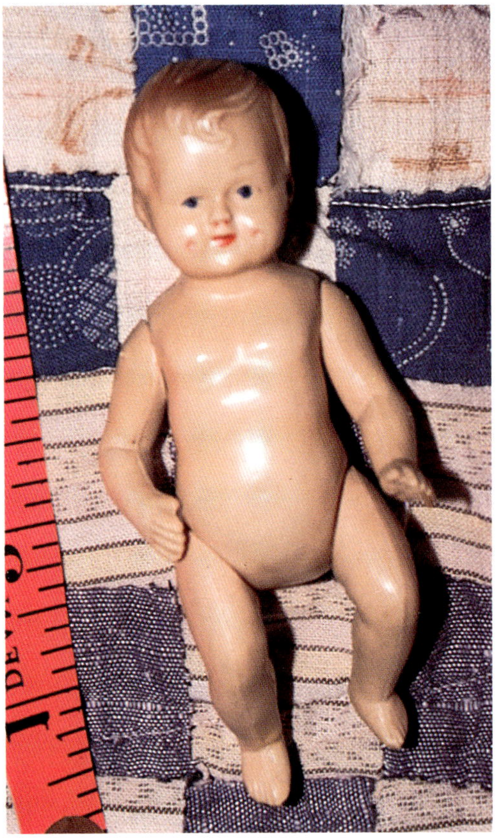

Ill. 192

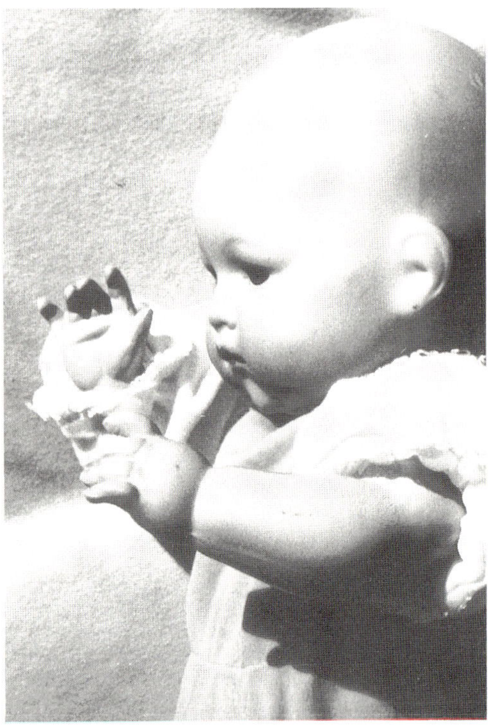

Ill. 193

(192) Small all celluloid baby, bought 1989. Mussel mark.

(193) Dream Baby with replaced Køge arms and hands. Benedicte Nielsen collection.

marked blonde wears a white trimmed pink cotton dress (Ill.187). Our triangle marked brownette twin wears a rose knitted dress and panties (Ill.188). That brownette is shown with another triangle marked blonde doll, both undressed to show body construction. The doll with brown hair has a fatter body and limbs (Ill. 189).

Our three thinner doll house dolls have identical construction, and their faces are miniatures of some larger dolls. One blonde doll has pearlized hair; the second shows traces of that hair coloring. Those two dolls have the two early Køge marks, the triangle and the enclosed mussel. The brownette has the enclosed mussel mark, and below the mussell marks is "16", probably a size number. The same number appears inside the hip sockets of all three dolls. They are posed to show articulation; one wears its knitted romper (Ill.190). Two dolls which inhabit an antique shop owner's doll house show the two different marks. The doll in green is marked with the triangle; the doll in red has the mussel. In appearance but not in dress or marks they are twins (Ill.191).

A 4 in. (10.2 cm) undressed baby with the plain mussel mark was an expensive flea market find in May 1989 (Ill.192). The celluloid dolls are much more scarce than we had noted six years earlier, and prices have escalated. The 1989 cost in D. Kroner was more than $10.00 per inch, comparable with the price of other small all celluloid dolls sold now in the U.S.

An incidental aspect of the celluloid doll industry was the manufacture of shoulder heads and also arms with hands. The Dream Baby of Benedicte Nielsen has replaced celluloid arms articulated only at the shoulders (Ill.193).

Summarizing our Køge findings, the Danish celluloid dolls are from 4 to 23 in (10.2 to 58.4 cm) tall (Ill.194). Identical body shapes have either the triangle Køge mark or the mussel mark. While some hair molds are identical, the dolls are not, because of the variety of hair colors: very dark brown, medium to light brown, medium to pale blonde, white, and the surprising and beautiful metallic gold and silver. Of more than 30 celluloid dolls documented, only eight have the unusual metallic hair. The most often seen hair styles are variations of molded waves and curls. Straight bobbed hair is seen on two doll house dolls. Only two wigged dolls have been found.

All the celluloid dolls made at Køge have an identifiable Danish look, although a few slightly resemble the German made Schildkröt dolls.

Articulation of the all celluloid dolls is only at the shoulders and hips, except those with socket necks. Of our dolls, only the tallest three, 18 to 23 in (45.7 to 58.4 cm), and only one wigged doll have socket necks. Numerous hand knitted or crocheted costumes suggest that the dolls may had been sold undressed as well as dressed. The dolls documented illustrate three of the four marks of the Køge dolls, those shown in Ill.152, 153, and 155.

Finding our first two dolls with two different marks called Køge had generated the research which followed. More Danish celluloid dolls are to be found in private collections than in museums. But the hunt and the research have not ended.

Ill. 194

(194) Group of undressed celluloid dolls, showing body construction and color variation

Chapter 9

DanDoll: A Modern Plastic Doll Firm

A chain of BR Toy stores has many outlets in the environs of Copenhagen. At several of those and at toy departments in the large stores on and near the Stroget, we found a wide selection of play dolls, stuffed animals, and Dam Trolls in 1989. But no dolls made in Denmark were found until we were directed to a BR shop near the east end of the Stroget. When we asked to see a doll made in Denmark, we were shown one brought from an office. The lovely doll (Ill.195) was labeled "DanDoll". ("Dan" is pronounced "dane".)

Correspondence with the firm in Hornsyld proved very encouraging. We made an appointment to visit the factory on our return to Denmark. An early morning drive from Billund through lovely countryside brought us to the plant on a hilltop at the edge of Hornsyld. Mr. Paul Tinglev-Hansen, the plant manager, and Mr. and Mrs. Einer Madsen, founders and owners of Dan-Hill-Plast A/S, were gracious in showing us the dolls in the display room and the factory, and in sharing the history of the company in large scrapbooks.

Dan-Hill-Plast consists of five divisions. DanFloat and DanFender make nautical products for boaters; DanTech makes rotational moulding. We are concerned here with DanCoin which produces delightful coin boxes we call penny banks, and toys for babies, and especially DanDoll which makes an extensive line of beautiful play dolls and doll size manikins.

In a third of a century this firm has progressively developed numerous exciting and new products. Animals and doll heads were designed by an artist from Aalborg, and molds and machines were first made by Mr. Madsen in 1957. Doll manufacture was begun in 1958; the first vinyl doll was Stella, 13 3/4 in (35 cm) tall. Then Mr. John Nissen, a sculptor of models and life sized manikins in his Copenhagen studio, Hindsgaul, became their artist. His molds of cats, dogs, hedgehogs, and other animals were used to manufacture soft vinyl toys, many adapted as coin banks, which increased the toy and doll lines of those two divisions.

By contract with Walt Disney Studios in 1960, production of toys and puppets of Mickey Mouse, Donald Duck, Lady and the Tramp, the Three Little Pigs, and others

Ill. 195

Ill. 197 a

Ill. 196

Ill. 197 b

(195) The first DanDoll found in Copenhagen. One of the "Quality Collection: Little Miss". Courtesy of BR.

(196) Anatomically correct boy and girl.

(197, a, b) "The World's Children". Photographed from Prospectus, courtesy of the Madsens.

Ill. 198 a

Ill. 198 b

Ill. 199 b

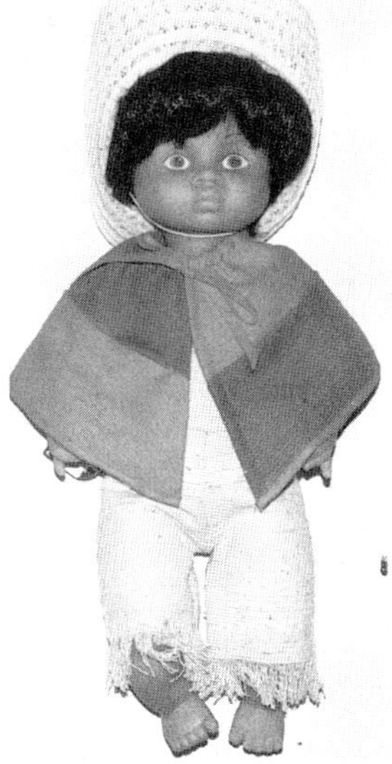

Ill. 199 a

(198 a, b) Oriental pair, a. full length; b. one in close-up.
(199 a, b) Pedro, a. full length, b. close-up.

was begun. In 1962 those characters were joined by the Flintstones, which became a popular display in the Copenhagen Forum.

In 1962 new rotation machines were needed; they were obtained from England. By 1965 the firm's production was being sold widely in Scandinavia and across the rest of Europe. Designs from Belgium were acquired in 1965. Anatomically correct dolls, introduced in France in 1967, were soon being made in many American and European doll factories including DanDoll at Hornsyld. In 1969 the widespread controversy over these brother and sister dolls was headlined in Denmark:

> Do you think that children should get sex information from the toy shop?

Child psychologists entered the discussion with their total approval, and the dolls are still being manufactured (Ill.196) and are popular. Another criticism about dolls was:

> "*Mindre engle; mere born*",

meaning the dolls should look "less angelic; more childlike".

Softer dolls were made in 1970, and twin dolls were featured. Czechoslovakia ordered 8,000 dolls in that year. Talking dolls were introduced in 1971. *Soren og Mette*, characters in a popular story for children, were made as dolls at that time. Plush toys soon followed and were most successfully marketed until about 1980 when stuffed toys from Korea became available at a much lower cost.

Expansion of the firm in the mid-seventies included a factory in Finland. It burned down, was rebuilt, and was sold to the Finns in 1976. Doll molds were bought in New York by the Hornsyld firm in 1975. A toddler doll with the tip of its tongue sticking out is one of the U.S. designs which is still being made. At the Nuremburg Toy Fair in 1978, "*arets dukke*", the "Doll of the year", was a very popular item. That doll was a best seller for the next four years, each year in a new costume.

Two new and very exciting series, the life sized baby dolls, and "The World's Children", were introduced in 1983. Ethnic pairs are very popular with all children, but American children prefer them in contemporary rather than national costumes, Mrs. Madsen told us. The six doll pairs represent Mexico, Pedro and Carmen; India, Krishna and Meera; the Orient, Ying and Zung; North America, Paw and Solsikke; Central Europe, Signe and Mads; and Africa, Sammy and Sarah. The two group pictures (Ill.197, a,b) are from the 1989 prospectus. Full length and close-up photographs show details of the Oriental pair (Ill.198, a, b) and the Mexican boy Pedro (Ill.199 a, b).

These dolls, 16 1/2 in (42 cm) tall, are beautifully designed and very durable. The sturdy cloth bodies are filled with solid foam forms. The ethnic dolls are guaranteed for two years, others for five. They are sold exclusively through distributors to kindergartens and nursery schools. Mrs. Madsen said that is why we did not see them for sale in the Copenhagen toy stores.

Ill. 200

Ill. 201

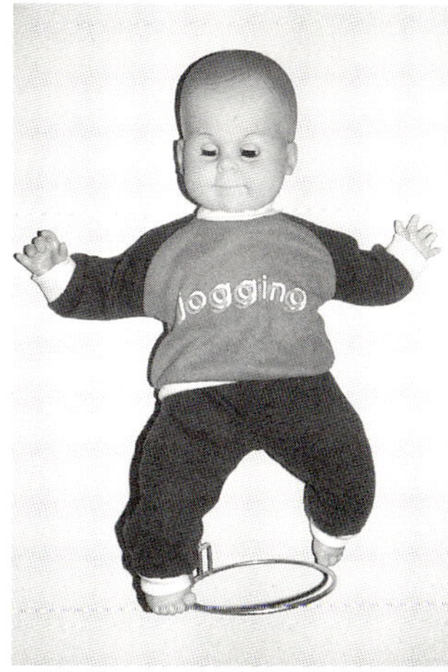

Ill. 202

Ill. 203

(200) Nigerian dolls. From Prospectus, courtesy of the Madsens.

(201) Doll of "Plaids and Checks".

(202) One of the "Little Imps".

(203) Dolls of "Crying Babies" from Prospectus, courtesy of the Madsens.

Ill. 204

(204) Dolls of "Outdoor" group from Prospectus, courtesy of the Madsens.

Another series of ethnic dolls in the 1989 prospectus is *Nigeriadukker*, the Nigerian Dolls. In this group there are four 15 3/4 in (40 cm) dolls (Ill.200). The bearded chief wears a toga and glass beads. The mother, carrying a babe on her back, wears a bandana matching a short dress and trousers of batik fabric. She is adorned with large hoop earrings, beads, and bracelets. The girl doll, similarly dressed, wears no headkerchief. The doll called Lagos wears a spectacular long dress and a draped turbaned headkerchief, and much jewelry, gold hoop earrings, and glass necklace, bracelets, and ankle bracelets. The batik fabric was ordered from the Netherlands exclusively for these dolls' costumes. The *Negerbørn* (black children) are anatomically correct babies, the girl about 4 3/8 in (11 cm) and the boy about 4 3/4 in (12 cm).

In the 1989 prospectus, 14 series of lovely play dolls are shown according to themes and harmonious costumes. The brown eyed cutie (Ill.201) is one of three 20 in (50 - 52 cm according to hair styling) dolls in "Plaids and Checks". The little boy in a jogging suit (Ill.202) represents the quintette called "Little Imps". Dolls whose pictures were reprinted from the catalogue portray "Crying Babies" (Ill.203), and "Outdoor" characters (Ill.204). All the dolls in this "DanDoll Quality Collection" have sleep eyes and lustrous curly wigs. All are guaranteed for five years. In beauty and durability these dolls rival any current play dolls, American or European made, which we have seen.

The soft vinyl squeak toys and the DanCoin items continue in popularity. Coin banks are now made in great numbers for Denmark's commercial banks, effectively advertising individual banks. Other coin banks are made for children in 20 animal shapes (Ill.205) and as vehicles, cars, buses and trains, with humorous faces painted on headlights and grills. The seal, the panda, and the elephant are three animal banks which represent endangered species. A percentage of the sales of the three banks is given to funds for those endangered animals. Our seal bank by DanCoin and our seal and whale banks by Dam Trolls were sent to our twelve year old grandson in California, who is very concerned for our endangered otters and whales.

A visit to the studio of Mr. John Nissen revealed that the hard plastic dolls of a series created in 1987 (Ill.206), "The Fashionables", are miniatures of his exquisite manikins sold universally. Listed in the 1989 prospectus, the manikin dolls represent eight decades of "Nostalgic looks — 1910: The Halcyon Days; 1920: Flappers Vogue; 1930: Thoroughly Modern; 1940: The New Look; 1950: Liberated Ways; 1960: Hot Pants and Long Legs; 1970: Jeans and Dreams; 1980: Laid Back and Lovely". Another dressed as a National Girl Guard is permanently displayed at the show room. That one was photographed with the 1920s "Jessie" in rose-pink, the one we brought home; it is a reminiscence of my older sister in her wedding dress. The two are shown in full length (Ill.207) and in close-up

Ill. 205

(205) Assorted banks called coin boxes with the Flintstones. The seal is one of three sold to benefit endangered species.

(Ill. 206) "The Fashionables" representing styles of eight decades, "Nostalgia looks". From Prospectus.
Left to right. 1910: The Halcyon Days; 1920: Flappers Vogue; 1930: Thoroughly Modern; 1940: The New Look; 1950: Liberated Ways; 1960: Hot Pants and Long Legs; 1970: Jeans and Dreams; 1980: Laid back and Lovely.

(Ill.208). Another interesting pair finished in silver or in gold are manikins for displaying jewelry and other accessories in fashionable boutiques. Arms removable at the shoulder sockets simplify dressing the manikins. All the costumes worn by these lovelies are designed and signed by Mrs. Else Madsen. Made in numbered series, they beautifully present 8 decades of fashion.

Nine of the latest creations, the new "United Children of our World", over 25 1/2 in (65 cm) tall, were completed for the 1990 Toy Fair at Nuremburg, Germany (Ill.209). The five beautiful models photographed in John Nissen's studio in 1989 were prototypes for children of Scandinavian or Northern European (Ill.210), Arabic or Indian (Ill.211), Oriental (Ill.212), Eskimo (Ill.213), and African (Ill.214) races. They very realistically and tenderly represent the children of those ethnic groups.

Clothing worn by these dolls was researched and designed by Else Madsen. With permission her photographs and quotations are used here. The fifteen dolls, 9 of the 1990 series and 6 additional dolls of 1991, are grouped here according to country or area. European countries (Ill.215,a-c) are represented, first by Heleen (a) wearing wooden shoes, a white cap and working dress typical of Volendam, Holland. Denmark is represented regionally by Karen (b) of Hedebo, in the "Harvest" dress decorated with many ribbons of silk, and Germany by Christel (c) of the Black Forest area bordered by Switzerland and France. Her costume is rich with braids and lace, and is recognized by the red pompoms on her hat.

Two children represent Greenland (216, d, e). The Eskimo boy Mika (d) "wears warm pants, a white, hooded parka and seal-skin *kamikker* (boots), to protect him from the terrible cold". The Greenland girl Ivalo (e) wears the special occasion dress "made from the finest Skins, Cloth and Thread with layers of beads and Pearls woven into intricate patterns".

The African continent is represented by three dolls (Ill.217, f-h). Muhammed (f) of Egypt works as a donkey driver in a small village near the Nile. He wears a striped coat called a *gelabil*. The Nigerian girl doll Tala (g) wears a typical bright colored, batik printed cotton dress with matching head dress. Glass beads decorate her neck, wrists and anklets. Fatou (h) of Ghana, a girl in the 1991 series, wears traditional cotton, an authentic wax print produced in a limited series. Around her neck are "amulets, to ward off ill spirits".

Salina of Bangladesh (Ill.218i) is a child who works as a chambermaid and attends school with the family's daughter. In the hot climate she "wears loose, comfortable clothes . . . *Salwar* (Pants) and *Kamez* (Blouse). . . of colorful printed cotton and on her head to protect her from the hot sun she wears a net *Chunny*."

The Orient is represented by Li Chun (j), a Chinese girl and Akane (k), a Japanese boy (Ill.219). Li Chun wears a winter costume of quilted jacket and pants, and snug quilted shoes. Akane, who goes to a fine school with

Ill. 208 a

Ill. 208 b

Ill. 207

(207) Fashionables doll of 1920s with National Girl Guard. Full length.
(208 a, b) Dolls in Ill.207; close-up.

Ill. 209

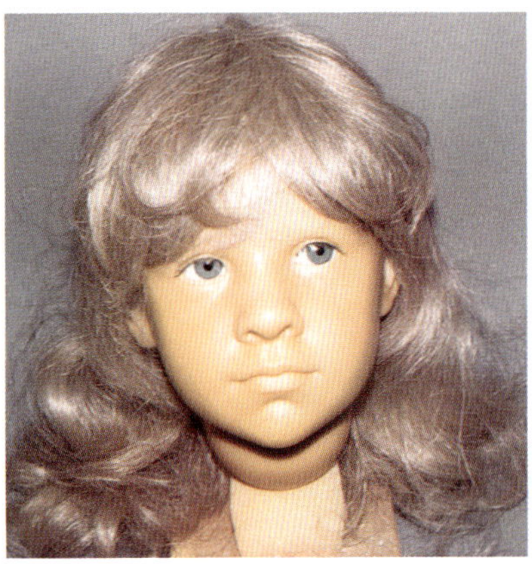

Ill. 210.

(209) New "United Children of our World" at Nuremburg Fair.
© Courtesy of the Madsens.
(210-214) "United Children of our World" prototypes of heads of dolls:
(210) Scandinavian child.
(211) Arabic child.
(212) Oriental child
(213) Eskimo child.
(214) African child.

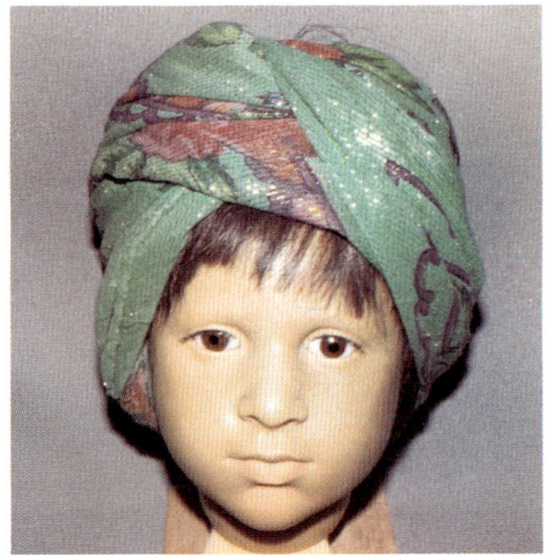

Ill. 211

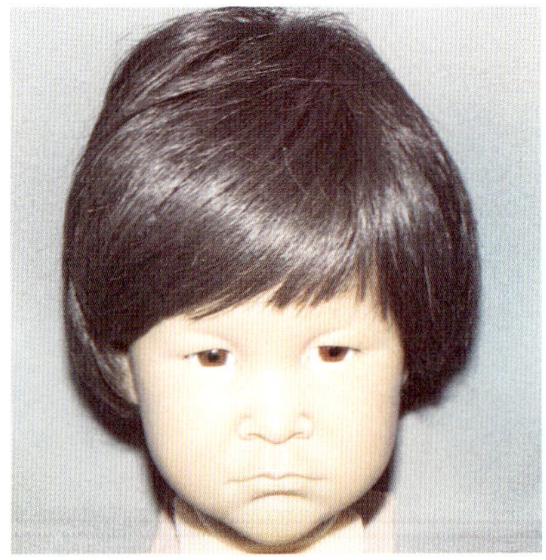

Ill. 212

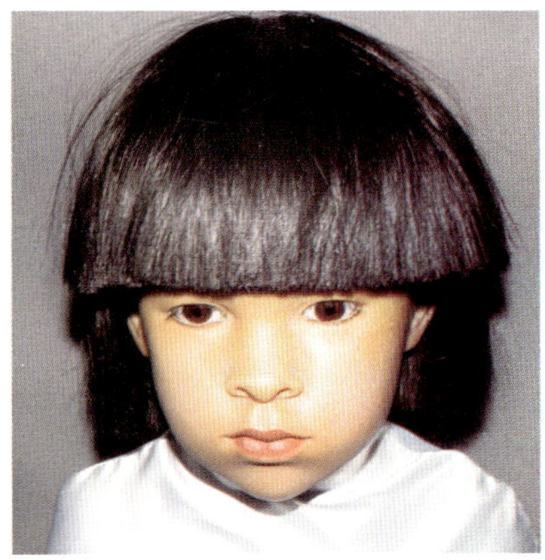

Ill. 213

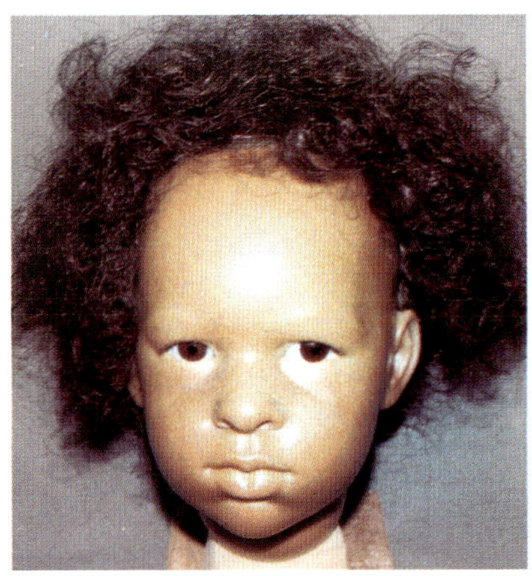

Ill. 214

aspirations of high school and university study, is dressed for his Judo class.

Philippini (l) is a beautiful young lady of the Philippine Islands (Ill.220), wearing "her 'Terno', a wonderful dress with Butterfly sleeves, which is woven out of Pineapple fibre".

Representing the western hemisphere is, first, Little Deer (m), a Red Indian wearing a buckskin dress and moccasins (Ill.221). The Mexican boy (n) Juan (Ill.222) lives in San Pedro in the Sierra Madre mountains, in a poor family whose "forefathers were kings of the Aztek Indians . . . hundreds of years ago". On his school days he wears typical mountain clothes including leather *zapatas* (sandals). Karina (o), a girl of Ecuador (Ill.223), also lives in the mountains, the Andes. In "a costume fashioned after her Inca ancestors, she wears a long, white chemise and a delicate blouse, embroidered with the mountain flowers of her homeland. A warm skirt is held in place by an embroidered belt and her sandals are brightly embroidered by matching flower trim".

All the costumes of these international children, some poor and others of better circumstances, reflect the Madsens' "insistence on excellence of design and quality". These beautiful ethnic dolls are made in limited editions of 1200 each, and are marked with serial numbers on the neck. The "United Children of our World" are distributed by Dynasty Dolls of Port Reading, NJ to shops in the U.S. We have heard from Hornsyld that Tala, the doll we admired in a doll shop here, is one of the best sellers.

Dolls representing international children are growing more popular in both the creations of doll artists and those made commercially. The Madsens' ethnic dolls have maintained that excellence which has not been surpassed by others. The fifteen dolls in the series "United Children of our World" are illustrated on the following three pages.

Ill. 215 a

Ill. 215 b

Ill. 215 c

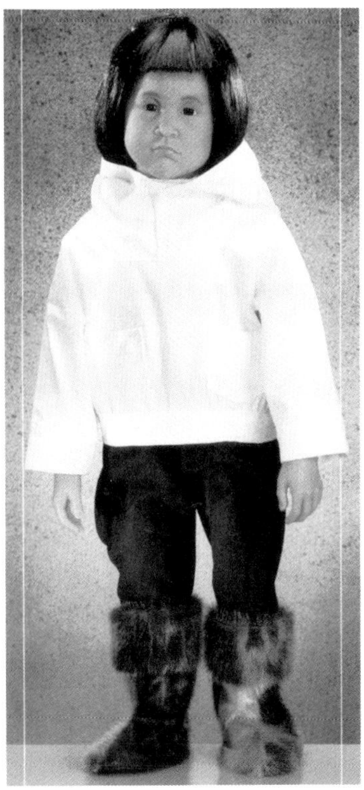

Ill. 216 d

Ill. 216 e

Ill. 217 f

(215-223) Dolls of series above according to areas, courtesy of Elsa Madsen:

Europe (215 a-c)
a. Heleen of Holland
b. Karen of Hedebo, Denmark
c. Christel of Germany

Greenland (216 d, e)
d. Mika, Eskimo boy
e. Ivalo of Uummannaq

Africa (217 f-h)
f. Muhammed of Egypt

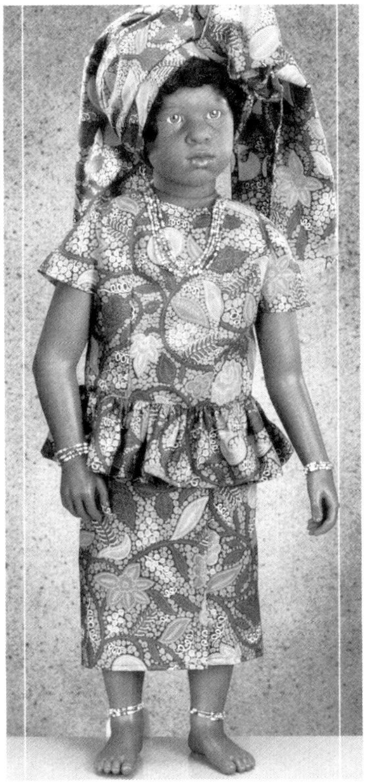

Ill. 217 g

Ill. 217 h

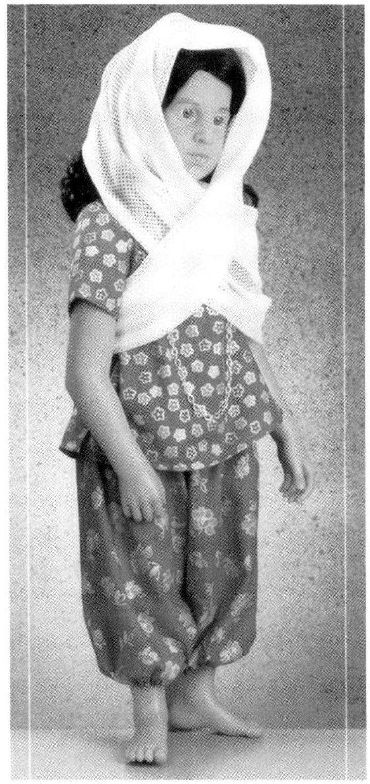

Ill. 218 i

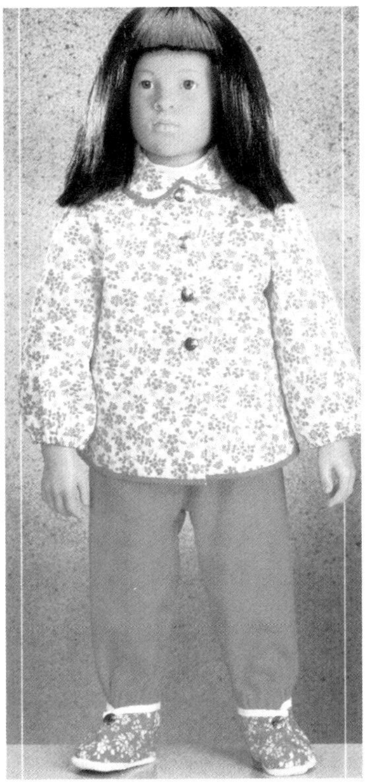

Ill. 219 j

Ill. 219 k

Ill. 220 l

Africa (217 g-h) Continued
g. Tala of Nigeria
h. Fatou of Ghana
Bangladesh (218 i)
i. Salina

The Orient (219 j, k)
j. Li Chun of China
k. Akane of Japan

The Philippines (220 l)
l. Philippini

Ill. 221 m

Ill. 222 n

Ill. 223 o

Western hemisphere (221 m-223 o)
m. Little Deer, Red Indian "of North and South America"
n. Juan of Mexico
o. Karina of Ecuador

All of the DanDolls are produced solely in Denmark. The yearly output is now 80,000 dolls. Countries to which they are routinely exported include Australia, Japan, South Korea, Canada and the U.S.A. as well as Continental Europe.

DanDolls are identified by grey plastic tags (Ill.224) which bear the logo of the firm, shown on the catalogue page, "The World's Children". This stylized logo is red and white, with the doll's arms bisecting the red field in the manner of one of the crossbars of the Danish flag. The logo is another example of the love of country expressed in the red and white of the Danish flag, as are the red and white costumes of the folk dolls labeled *Mor Danmark* (Mother Denmark).

Ill. 224

(224) Stylized plastic hangtag on DanDolls.

Chapter 10

More Contemporary Artist Dolls

Artist designed and molded historical figures of porcelain currently made by Kirsten Gynther Eriksen in Funin were the content of Chapter 2. Pamela Nedergaard's skillfully created needle sculpted dolls appeared in Ch.5 and in Ch.6. For her first doll of cloth she molded the head of Cernit (Ill.225). Other molded or sculpted dolls are being made in their homes by two talented artists who live in Jutland.

Ella Hass, an artist in Torring, Jutland, creates realistic baby and toddler dolls. She sculpts the individual faces inspired by her own children, and completes the entire dolls with cloth bodies. Mrs. Hass was not available for interviewing, photographing or purchasing her dolls. They had been advertised in the U.S. in 1989.

Another Jutland artist making unique dolls and puppets is Astrid Bernth. Astrid was born in Switzerland and studied arts and crafts in a Basil crafts school. Her dolls were exhibited in Switzerland in 1978 and 1981; since 1985 her work has been exhibited extensively in her city of Silkeborg. A 1987 "By Invitation Only" exhibition in Aarhus included Astrid's dolls. She has taught many doll making classes; she also volunteers her time and artistry to numerous schools.

Astrid's dolls are her own designs, technique, and formula. She perfected her simplified stick puppets to teach to school children through grade ten. The head is made by mounting a little wooden ball atop a dowel stick and molding clay features on that. Only small pieces of fabric are needed for the limbs and costume, something a pupil can accomplish in a limited time. Her own stick puppets have sculpted heads and more elaborate costuming. She deftly molded and finished a puppet head at the Euro Toy Show and Sale in Billund in October 1989 (Ill.226). Fascinated observers learned much about her sculpting talent, and all were amused with the characters and caricatures so adroitly created.

The modeling substance she uses looks like putty; it is "a modeling clay of wood powder and glue. In this clay, bits of material of many kinds are important elements emphasizing the character of my dolls", she said at one of our interviews at the

Ill. 225

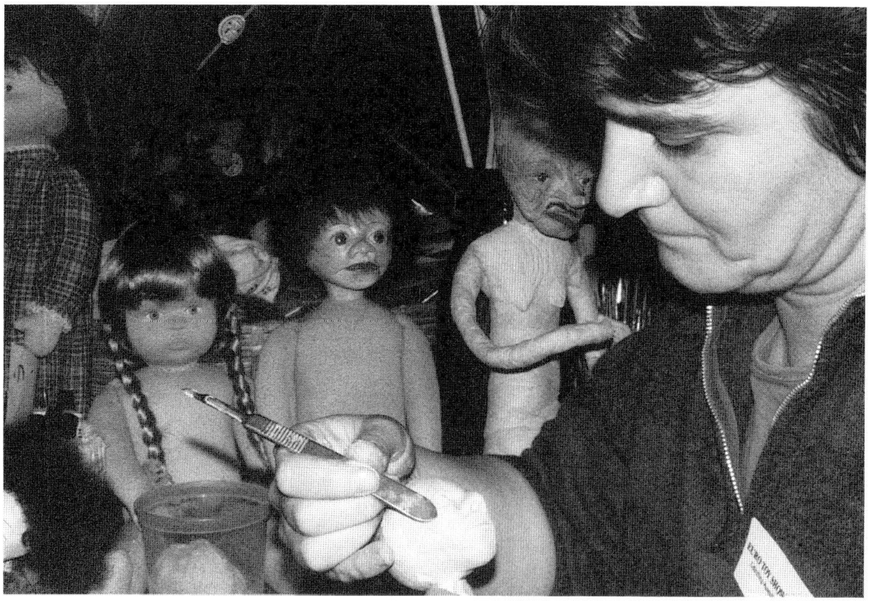

Ill. 226

Ill. 227

(225) Pamela Nedergaard's Cernit-molded doll.
(226) Astrid Bernth with molded puppet.
(227) Dolls and puppets in A. Bernth's booth.

Ill. 228

October show. The artist's imagination is displayed in the dolls pictured in her booth there (Ill.227), representing the "Bad Boy" in the upper left, the laughing clown in the lower right, and four others of assorted expressions and costumes. Part of a stick puppet appears near the upper left corner. Her Bad Boy is shown in a close-up (Ill.228).

She summarized her creative philosophy in her presentation at the gathering of doll artists at our hotel later, "With my dolls I tell about people I meet and about fantasy figures". The green eyed doll "Anne Marie" (Ill.229) represents a farmer's daughter dressed to go to town for a first time. That doll's red hair is fox fur. Another doll has a wig of fleece of a merino sheep (Ill.230); others have human hair wigs.

Astrid Bernth's one of a kind dolls, which in 1989 cost 2000 to 2400 Danish Kroner, over $300.00 each, are museum worthy, and would be the pride of any collection of contemporary artist dolls!

These needle sculpted, clay sculpted, and individually molded dolls by today's artists are impressive additions to the dolls made in Denmark.

Ill. 229

(228) Bad Boy by A. Bernth.
(229) Green eyed Anne Marie by A. Bernth.

Ill. 230

(230) Dolls by Bernth
a. with wig of human hair (in Ill.228)
b. wig of merino sheep fleece by A. Bernth.

Chapter 11

Mystery Dolls; Miscellaneous Dolls

This chapter will include a series of dolls about which not much information was available, dolls according to the materials from which they were made, and a few other specific kinds such as mascot dolls. Some dolls have been dated by the museums, by the collectors, or from advertisements. Some for which neither dates nor makers have been identified remain mysteries. Help from readers will be welcomed.

Wooden Dolls

Few very old Danish wooden dolls have been found; those documented were photographed in the doll collection at Legoland A/S. The wooden doll named "Bedste" (Ill.231) is dated ca 1860.

This plain looking doll, about 6 3/4 in (17 cm), is dressed in cotton. With a copper and wood coffee mill on her lap she is appropriately placed in a kitchen setting. Another doll (Ill.232, 233) has well carved facial features characteristic of skillfully made china or bisque head dolls of the late 1800s, wavy hair, and well defined eyes and mouth. Where the end of her nose has been broken off the wood grain shows clearly. The natural wood of her face, with no added color, is part of her charm. This doll, dated 1865, is 21 1/4 in (54 cm) tall. She wears a splendid costume, a dark colored dress, an embroidered shawl, an ornate brooch, and a fancy bonnet tied with a long silk scarf.

No wooden dolls to illustrate the decades following, until two doll house dolls (Ill.234, 235), whose costumes suggest the 1920s, were shown to us in Copenhagen. The black doll wears a large hair bow and a dress with cape styled sleeves. The white doll's straight dress has a scarf sized collar tied in front. Her shoes are painted red; the black doll's shoes are unpainted. These all wooden dolls are articulated only at the shoulders. The hands are paddle shaped. The black hair and facial features are painted on what appear to be lathe turned heads. Their dark brown eyes, turned flirtatiously to the side, can be described as googly. Adding to their flirty appearance are the painted heart shaped mouths, partially worn off the white doll.

Ill. 231

Ill. 232

Ill. 233

Ill. 234

Ill. 235

Ill. 236

(231) Wooden head doll Bedste, courtesy of Kirsten Martin Stadelhofer, Don Luckey photo.

(232) Carved wooden doll head, courtesy of Kirsten Martin Stadelhofer, Legoland A/S.

(233) Close-up of doll in Ill.232, courtesy of Kirsten Martin Stadelhofer, Legoland A/S.

(234, 235) Pair of wooden dolls, front and back views. Courtesy of I. and P. Pedersen.

(236) Fortune telling doll with papier mâché head, courtesy of Kirsten Martin Stadelhofer, Legoland A/S.

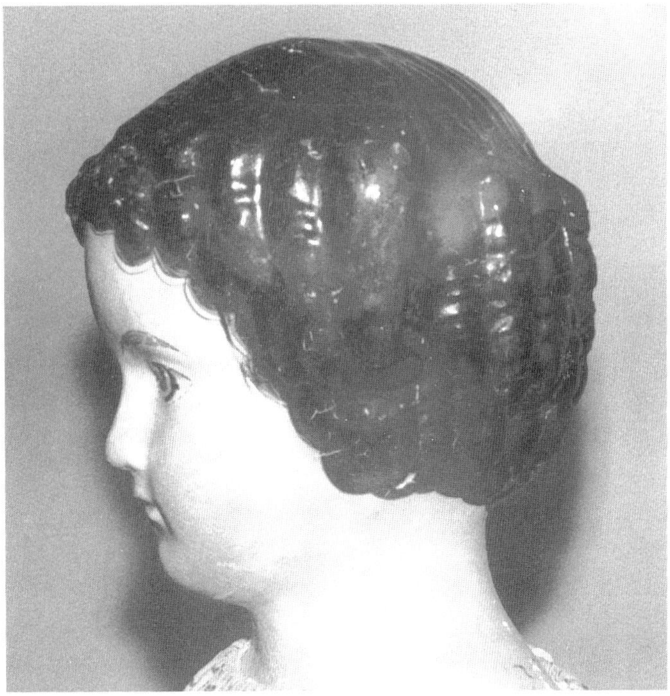

Ill. 237

Ill. 238

Ill. 239

Ill. 240

(237) Old papier mâché doll, courtesy of Kirsten Martin Stadelhofer, Legoland A/S,

(238, 239) Papier mâché mystery doll. Close-up of head of mystery doll.

(240) Mark on mystery doll. Sketch by Eileen van Baren.

It is quite possible that other wooden dolls were made on lathes in Denmark as well as in England and other countries. Little wooden head folk dolls and Nisser made by Pamela Nedergaard were presented earlier. The jumping jack, a doll-like toy called the *sprællemand* in Danish, is currently made of painted wood. These toys will be documented later with their Danish and other history.

Papier Mâché

Dolls of papier mâché were made in Nuremberg and Sonneberg as early as the eighteenth century. Waste materials from book making in the Sonneberg area were plentiful, and dolls were inexpensively made by molding a mass of paper scraps with some kind of binding material. Durability was lacking in these doll heads, and few very early ones are found.

A Danish doll about 9 1/2 in (24 cm), dated ca 1830, was photographed at Legoland A/S. This *spadukke*, (fortune telling doll) whose underskirt consists of folded papers with fortunes written on them (Ill.236), has a papier mâché head which resembles lathe turned wood. Her arms, legs and body are wood.

She is listed here instead of with wooden dolls because the material of a doll's head, not the body, determines its classification. Fortune telling dolls were only occasionally made in Europe; we were pleased to have found this one in the museum.

Other beautiful examples of papier mâché heads in the doll museum at Legoland A/S are on dolls named "Jeanina", ca 1915, and "Angelica", ca 1860. Angelica presents a three-way look, with painted hair or with a dark wig, and with a bonnet [68]. Another lovely papier mâché doll head with painted dark curls was photographed there (Ill.237). The dolls Jeanina and Angelica have recently been identified by Kirsten Martin Stadelhofer as German dolls. Is the one shown in 237 Danish?

An 18 in (45.7 cm) doll resembling a German bisque toddler (Ill.238, 239) was a Copenhagen antique shop find in 1978. The seller could tell us nothing about its origin. We thought it might be a German doll. A description of the doll, its mark, and photographs were sent to a German author and former editor of *Puppen und Spielzeug* (*Dolls and Toys*). She translated and published the article as a mystery doll; "Who is it?" [69]. Not a clue was sent from her readership to either her magazine or to me in Missouri. The puzzle deepened when we saw another doll like that toddler in the Israels Plads flea market 11 years later. Again the seller could not answer our questions about the doll's origin. The appearance and mark were those of our mystery doll bought earlier; on the back of both dolls' necks appears "CHK//50CM" (Ill.240). Use of the German measure cm adds to the puzzle since Denmark used the measure *tomme* and now uses inches.

A clue might be apparent in Finn Risom's discussion in Ch.7, about the use of molds made of dolls or doll heads from other countries. Might the K in the mark suggest Køge, where dolls of papier mâché were made during the 1940s and 1950s? Having seen dolls with the CHK mark only in Denmark strongly suggests that they could have been Danish made. Museum authorities could give no clues.

In October 1989 we found a papier mâché doll head with a curly mohair wig marked "Made in Germany//U.S. Zone". This is additional evidence of that very early product, papier mâché, having been used for doll heads as recently as the years following WW II.

Our doll's papier mâché head is on a papier mâché body which was occasionally used with a Danish made composition head. Her blue sleep eyes, open-closed mouth with two upper teeth, brown human hair wig, and the red tracings on fingers and toes are details of well made dolls.

The question regarding the K mark arises again about the papier mâché head on the doll Haldora, detailed in Chapter 3. High on the back of the head is the mark "8/O"; "K" is marked lower on the neck. The letter K suggests the possibility that this doll was also made in Køge. In a great grandmother's trunk for many years, the doll is undated.

We have found no Danish made doll house dolls of papier mâché, but cannot say they didn't exist.

Metal Dolls

In the long list of materials from which early dolls were made, Sisgaard and Varnild include iron, brass and other metals [70]. The countries which made them are not specified, but one might assume that unmarked metal dolls were Danish made, as were many unmarked composition dolls.

An all metal doll, shown wearing an original Danish folk costume and also undressed, purchased in 1949 [71] might be Danish. This unmarked 5 in (12.7 cm) doll, articulated at shoulders and hips, shows strong facial resemblance to some of the all celluloid dolls in Chapter 8.

Two dolls described next were bought in 1989. The 7 1/2 in (19.1 cm) all metal doll (Ill.241) is articulated at shoulders and hips. His brown wavy hair and deep blue eyes show a similarity to some of the Danish celluloid dolls which slightly resembled the German Schildkröt dolls. In spite of this doll's rough use, it is evident that care had been taken in the finishing. The facial details, especially the eyes, eyelashes and nostrils, had been well painted. Fingernails and toenails were outlined in red. No marks exist; he cannot be dated. His little shirt and trousers are obvious replacements.

Ill. 241

Ill. 242

(241) All metal doll, unmarked; thought to be Danish.
(242) Bisque head man doll. Photo courtesy of a friend.

Ill. 243

Ill. 244

Ill. 245

Ill. 246

(243-245) Three reproduction bisque head doll house dolls. Photos courtesy of Camme Bache.

(246) Mark on dolls by Camme Bache. Eileen van Baren sketch.

The second doll, thought to be of metal when we bought it, is the Kaiser Baby look-alike described in Chapter 7, Ill.142. Most interesting about this composition-classified doll is the unmistakable material, a metal, of the bent legs. The left leg has been cracked entirely around the knee; a composition limb would probably have been severed in such a break. An examination at cracks and chipping through the finish of the doll's head and arms reveals a composition material unlike the metal of the legs. The clean cloth body, apparently original, indicates that this little baby doll was probably assembled with composition head and arms, and the unusual metal legs.

Bisque Dolls

We have found only one original doll which is called a "clay boy" (a type of bisque), and the Tivoli clown presented later which might have been made in Denmark. The boy doll of almost 17 in (43 cm) is a single entry in the May 1990 *International Legetojsaucktion (Toy auction) catalog 596*. It is identified in entry 84 as "probably 'Sorring' Denmark ca 1940." The description is "Clay-doll, boy with painted clay-head, closed mouth and painted blue eyes...Clay body..." In Stuttgart, Germany a doll of clay-like bisque called *Keramic-biscuit* (ceramic-bisque) had identified our Armand Marseille baby 2549. Our baby doll is classified as "painted bisque" in an American book, and no category to compare with the coarser ceramic bisque seems to be used. The boy doll is articulated at the shoulders and hips; his molding resembles that of a smaller all metal boy described in this chapter (Ill.241). The Sorring boy wears a 2-piece sailor suit with short pants buttoned to the shirt, short sox, and leather shoes.

An exciting series of reproduction doll house dolls with bisque heads and limbs can be included here. We had bought a doll house man at Legoland A/S in 1979 for a friend (Ill.242). Excitement grew when that doll was positively identified in an advertisement from I. and P. Pedersen, titled *C.B.D.K. Dukker*, which listed an address and a telephone number in Praestø.

Camme Bache of that firm has kindly sent information and photographs of some of her *dukkestuedukker* (doll house dolls). Her struggle to learn doll making and finding needed materials will be included in Chapter 12. She made and sold twelve complete dolls [72]. Ten taller 6 1/16 in (15 cm) dolls with numbers are:

202 B. Lady in blue dress	207 K. Cook
202 S. Parlor maid	301 H. Man in leisure jacket
203 R. Lady in red dress	302 S. Man in sports clothes
203 N. Nurse	305 C. Man in military uniform
204 U. Young girl	306 F. Man in formal wear.

The following pair is 3 15/16 in (10 cm):

402 R. Little girl and 501 H. Little boy.

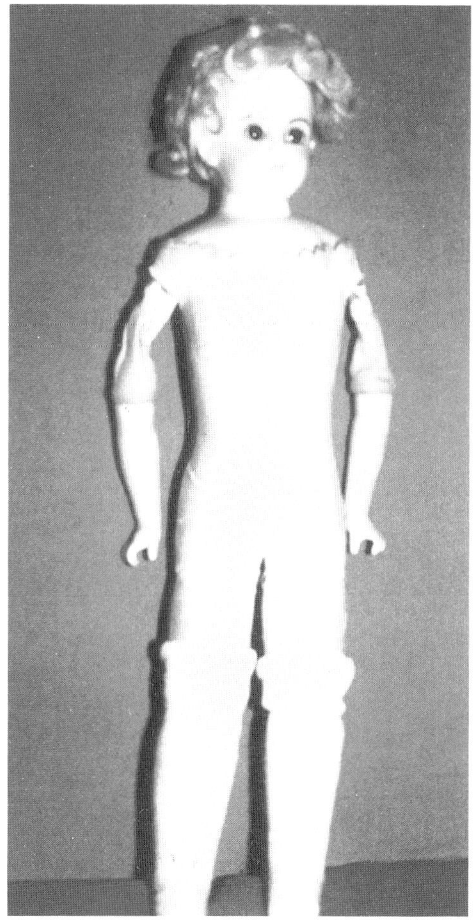

Ill. 247

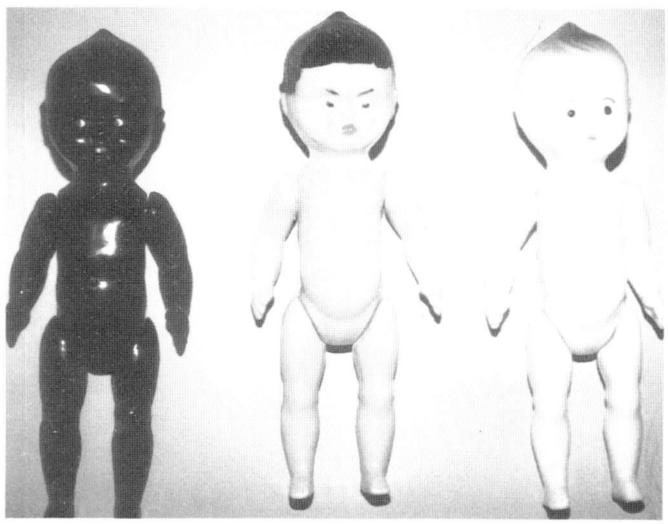

Ill. 248

(247) Reproduction Kestner bisque doll on leather body, courtesy of M. Siesbye.

(248) Three plastic doll house dolls, courtesy of Kirsten Martin Stadelhofer.

Ill. 249

Ill. 250

Photographs of dolls (Ill.243 - 245) sent from Praestø are identified from the above list as the young girl, the man dressed in sports clothes, and the man dressed in formal wear. Two heads, those of a man and a lady, and arms and legs were sold in kits, available when we bought the little man, Ill.242, at Legoland A/S.

The dolls are all marked with Ill.246; some may have the initials of her painters, L.H. for Lilian Hansen or H.T. for Helga Torp.

Camme Bache wrote that she had made all the doll house doll molds from German dolls; there were no original bisque doll house dolls made in Denmark. "I think I made about 2,000 dolls (since 1978), mostly dollhouse dolls. I also made moulds, German hands and Kling heads" [73]. Those she molded from her collection of antique Kling dolls, her favorites.

Other reproduction bisque dolls made in quantity for sale are seldom found. One example is the 3 1/2 in (8.9 cm) reproduction Bye Lo Baby sold as a souvenir doll at Mrs. Margot Siesbye's museum in Copenhagen. Beautiful Kestner dolls (Ill. 247) were made for their own pleasure by Mrs. Siesbye and her young receptionist Kate. Other doll lovers have shared, for documentation here, bisque reproduction dolls made in classes and in their homes. Mrs. Ester Pedersen's bisque 3-faced doll will be presented in Chapter 12. Majbrit Muzs of Charlottenlund made a lovely S.F.B.J. doll in a class. The challenge of making and dressing beautiful bisque reproduction dolls is growing in Denmark as well as in England, Germany, and the Netherlands.

Hard and Soft Plastic Dolls

Kirsten Martin Stadelhofer at Legoland A/S graciously provided her three small hard plastic dolls (Ill.248) for documentation here. These 4 1/2 in (11.5 cm) dolls were produced and sold as Lego toys from 1949 to 1956. Made from the same mold, they are articulated at the shoulders and hips. Their colors and painted hair and features represent three races. The Oriental doll's eyes slant; the black doll's eyes are blue, a rare phenomenon [74].

The little dolls were no doubt the forerunners of the tiny dolls in the LEGO toy kits to be assembled. There are small articulated figures in kits of those purchased in 1990, the Town System, the Space System, and two labeled the Castle System. The 1 5/8 in (4.2 cm) figure of Black Monarch's Ghost in one Castle System has the unique feature of glowing after being exposed to the warmth of a light bulb (Ill.249). Other articulated figures and animals, machinery and tools, and space gear are shown on the backs of the kits' packages.

Sigsgaard and Varnild reported that by 1950 several Danish firms had begun making plastic dolls. One company cited is *Nordjysk Plastic-Industri* (North Jutland Plastics Industry) [75].

Soft vinyl plastic dolls have been dated 1950 and 1970, in advertisements shared by I. and P. Pedersen. Tekno, a Danish toy firm begun in 1928, was producing construction sets of metal from 1936 to 1940 [76]. In a 1950 toy journal [77] Tekno had presented their new line of dolls which they named "*Tomte Dukker*": "We call your attention to the good and beautiful Tomte Dolls of vinyl." (The translater told me that *Tomte* is the Swedish word for Nisser.) Dolls listed in this announcement were Little Girl, Bonny, Johny, Jane and Andy; retail prices listed were from 27.25 to 5.50 D.Kr.

Johny and Jane (Ill.250), Nos. 704 and 705, were priced at 12.50 D.Kr. each, less than $2.00. Also listed in the advertisement were trains and boats, Teddy and Bamse bears, jumbo and little elephants, a dog and a clown, priced under $1.00 each. A cradle for a baby doll cost 15.50 D. Kr, about $2.25.

It was not unusual for a factory making dolls of vinyl to produce a variety of other playthings such as those listed here. That trend is apparent in the Dan-Hill-Plast firm's production.

A second advertisement ca 1970 identifies the 6 in (14.9 cm) articulated black baby doll (Ill.251) found in Copenhagen in 1989. A firm at Valby was named *J. Kofods Fabriker A/S*. The trademark in the advertisement and on the doll's back is the familiar Schildkröt turtle in the diamond (Ill.252). "*Eneret*", exclusive rights to the trademark's use by the Kofod firm, are cited in the advertisement. A brown vinyl doll named *Pummelchen* appears in a 1961 Schildkröt advertisement reproduced in a 1986 German doll book [78]. That doll is listed with several others as "*typische Kinder der Vinylzeit*" ("typical children of the vinyl period"). Two styles, with black painted hair and with an Afro wig, are shown. Except for its lack of hair color, our little black doll looks like the German made vinyl dolls.

The Schildkröt/Kofod connection of the 1960s had been a mystery. No response was given to a telephone number listed in the advertisement, and all correspondence to the firm had been returned stamped "address unknown". My conclusion that this firm probably no longer exists was confirmed recently by Marie Bach. She clarified the involvement of Schildkröt with Kofod, cited in Chapter 8.

Vinyl doll production continues today, as seen in the DanDoll story in Chapter 9 and in a later entry of this chapter, Mascot and Advertising dolls.

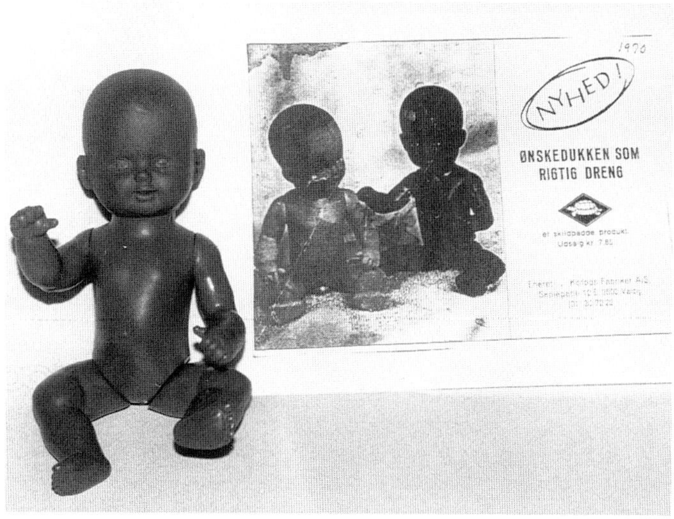

Ill. 251

Ill. 252

(251) Black doll with advertisement showing the Schildkröt connection; advertisement courtesy of I. and P. Petersen.

(252) Mark on Schildkröt doll. Eileen van Baren sketch.

Ill. 253

Ill. 254

(253) *Spraellaemaend*, Knold and Tot, photographed courtesy of *Dukkemuseum* in Ebeltoft.

(254) Papier mâché *Spraellemand*, courtesy of Kirsten Gynther Eriksen.

Spraellemaend, The Jumping Jacks

The jumping jack, a doll-like toy called *spraellemand*, has a very early history [79] and is included because it remains a popular toy in Denmark. Its early name *Pantin* came from the Paris suburb of the same name, where it was popular from 1746 to 1756. In Germany the toy is called the *Hampelman*; plural is *Hampelmänner*. In the 1700s the toys had also been made in Oberammergau; very old examples are exhibited in that city's museum and are being replicated by today's wood workers there. Many had appeared at early Nuremburg Toy Fairs. Through the next two centuries numerous characters and caricatures were popular. Both the German and the French toys were made as lithographed sheets which required cutting out and assembling.

In Denmark as early as 1890 a series of *spraellemaend* was made representing characters of Tivoli's panto-mime theater. Those included Perriot, Harlequin and Columbine. Later Kewpies and the roguish pair, *Knold and Tot*, were made (80]. A framed uncut sheet of the boys (Ill.253) is exhibited at the Ebeltoft *Dukkemuseum*. That pair was the work of Alfred Jacobsen. About fifty years later another Danish artist, Adolph Holst in Aalborg, made caricatures of the despised von Ribbentrop, Hitler, Goering and Goebbels [81].

Other early evidence of this popular toy in Denmark [82] appeared in the Dec.1, 1894 issue of the magazine "*Husvennen*" ("The Household Friend"). On the cover was pictured a young girl with a *spraellamand*. A four stanza poem in the magazine described the sorrow of the poor little girl who desperately wished for one toy, a *spraellemand*, of her very own.

A wonderful *spraellemand* Nisse (Ill.254) was photographed in the home of Kirsten Gynther Eriksen. That jolly cardboard fellow has the mark of the Royal Copenhagen firm on the back of the *abelkskiver* pan in his left hand. Mrs. Eriksen has consulted the archives of the firm, and learned that no *Spraellemand* had ever been made there; the trade mark was pirated. She dated this toy ca 1941. Now she is creating jumping jacks, 9 in (23 cm), of this design as toys, and smaller ones as Christmas tree decorations.

A series of contemporary wooden jumping jacks (Ill.255) was bought in Copenhagen in 1989. Numerous shops sold these characters and clowns, Vikings, and others made in Denmark. The popular fisherman, the chimney sweep and the postman, a child in Greenland dress, and a mermaid caricature by Larsen measure ca 3 1/4 in (9.1 cm). A Santa measures 4 in (10.2 cm). They are well designed and finished; both fronts and backs are painted in detail on all but the Santa. These simple little wooden novelty toys will remain popular Danish souvenirs.

Mascot Dolls and Advertsing Dolls

Other contemporary Danish novelties included here are dolls designed as mascots for sport events. A mascot doll, Troligan, is the first illustration in Chapter 4. This troll doll was made for the European championship football game Denmark played in France in 1984. Ten thousand were sold before the game; none sold after the Danish team lost, Niels Dam informed us.

Another mascot doll was photographed at the Israels Plads flea market in 1989. It wears a horned red hat and a red shirt plainly marked "D B U//Carlsberg". The three initials stand for the *Danske Baldspil Union* (Danish Ballgame Union). The uniform includes white cotton shorts, red shin guards, and black hard vinyl shoes with cleats on the soles.

Carlsberg Breweries had sponsored a competition for a mascot doll honoring the 1986 world championship football match in Mexico. "The winning mascot was made by Eilif Frederiksen in Olstykke, and we named it CARLO after the mineral water that we launched as a world championship drink. The mascot...about 20 cm long (8 in) was only made in a very limited number..." [83]

A Carlo doll was sent to us by Helle Madsen in December 1990. This hard vinyl plastic doll is articulated at the neck, shoulders, and the left ankle, and has its own football (soccer ball). The Carlo doll is labeled the "*VM Fodbold Landsholdets MASKOT*" (the VM Football League's Mascot). The manufacturer is *Bording Formulartryk A/S*, at Odder and at Herlev [84].

The costume is by the Garage Clothing Company [85]. The mineral water label sent from Carlsberg Breweries is shown with Carlo, with the kind permission of Helle Madsen (Ill.256).

Identification of these two purposeful dolls suggests that other mascot dolls might be made honoring special Danish sport events. However, Carlsberg has no plans for similar competitions, Helle Madsen advised.

An unmarked doll sold in Copenhagen as the Tivoli clown in 1989 might be considered a mascot of the famed amusement park there. It is 8 1/2 in (21.7 cm) tall including the cap; the molded bisque porcelain head has painted dark blonde hair, dark brown eyes with highlights, and a light blue star on the left cheek. The doll's body with wired arms and legs allows posing. The costume is red and white (Ill.257). The Tivoli clowns were sold in hotel gift shops and many kiosks; no one could identify their source.

These mascot dolls, especially the Carlsberg doll and the Tivoli clown, would probably be considered advertising dolls by collectors of that specialty in the U.S.A. Although we had first thought Carlo was an advertising doll, we have found no dolls in Denmark which are made available with required proofs of

Ill. 255

Ill. 256

(255) Danish jumping jacks, right from center top: postman, fisherman, chimney sweep, Santa, mermaid, and Greenland child.

(256) Carlo, for Carlsberg competition. Courtesy of Halle Madsen.

Ill. 257

(257) Tivoli clown, 1989.

purchase, the usual manner of acquiring many advertising dolls here.

A few more dolls which might be considered for either category — mascot or advertising dolls — will be described here. Lego plastic bricks, made since 1949, have been used to construct attractive doll-like figures seen at Legoland Park, in the sales room (Ill.258) and elsewhere. In the adjoining hotel a trio of Nisser in classic red and grey (Ill.259) and more doll figures are reflected in mirrors. The modernistic attractions could be considered advertising dolls; their construction is obvious.

Another Lego brick construction (Ill.260) identifies one of several BR toy stores on the *Stroget* in Copenhagen. And a large plastic figure of the BR trademark, a Danish guard (Ill.261), is seen among several others on posters at a shop's entrance there. Of the 28 BR toy shops listed in the Yellow Pages of the 1989 Copenhagen telephone directory, we investigated five. All their personnel were considerate and helpful during our modern doll hunts. It was in one BR store on the *Stroget* that we learned about the DanDolls.

Dolls of different materials and for different purposes have been presented in this chapter, with dating and documentation of as many as possible. However, incomplete information causes others to remain in the category of mystery dolls.

While it will be argued that some of those in the last two categories are toys rather than dolls, they are related, and, of course, become a part of the history of dolls and of Denmark's athletics, industry, and recreation.

ADDENDA:

The Danish made doll from Villy Nielsen APS was introduced on May 8, 1992 at F.A.O. Schwarz in New York. This 11 1/2 in Judith "Mommy-to-be" doll has an anatomically correct baby in her "removable tummy". The controversial presentation by the U.S. press suggests unpredictable acceptance of this doll.

Ill. 258

Ill. 260

Ill. 259

Ill. 261

(258, 259) Dolls made of Lego bricks, could be considered advertising dolls. Photographed at Legoland A/S, courtesy of Kirsten Martin Stadelhofer.

(260) BR toy shop logo of Lego bricks.

(261) BR logo made of plastic, in front of entrance.

Chapter 12

Today's Doll Artists and Artisans

Much of the excitement concerning Danish dolls today is due to the talents and ambitions of those individuals whom the author and photographer have met, and no doubt, others not yet discovered. These creative people are designers, commercial artists, sculptors, stylists, teachers, and also artisans or craftsmen who skillfully make dolls using molds of existing dolls or commercially produced molds.

Some of the artists are self taught. Some have developed their artistry in an attempt to salvage or replace a cherished childhood doll. For one senior member of this group, doll making began as therapy during a long convalescence. All of them command our respect, our admiration. Each will be presented in the sequence in which the reader has met her or him.

Kirsten Gynther Eriksen

Kirsten Gynther Eriksen knew heartbreak when her best loved childhood doll was broken beyond repair. A teacher of gymnastics, wife, and mother of sons, she determined to remedy both breaks, though she had no artistic training. She began reading and experimenting. The creation of each doll was surely a trial-and-error process. She had no teacher to guide her in designing the figures, making molds, or drafting patterns for the bodies and the exquisitely accurate costumes. Photographs taken in the workshop show some of the activity involved in Kirsten's doll making (Ill.262, a - d).

These dolls are more than figurines. The well designed heads and hands, legs and feet are attached to cloth bodies and arms. Positioning of the arms is possible because of wires inside them. For the men dolls a unique mold for porcelain from the waist including the feet provides stability, and allows easier costuming of the legs in short breeches and boots.

Ill. 262 a

Ill. 262 b

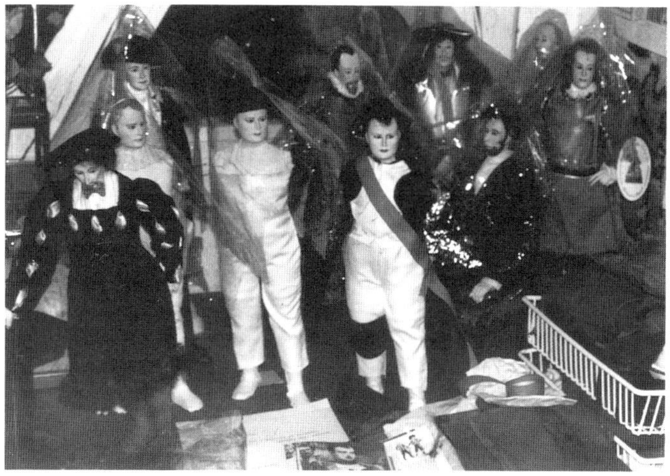

Ill. 262 c

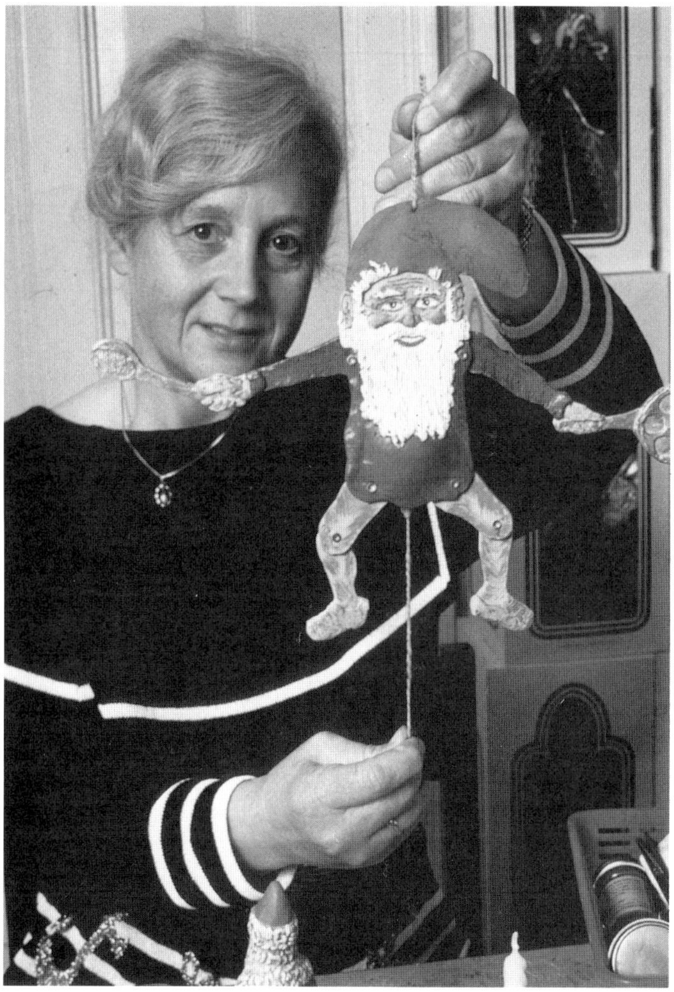

Ill. 262 d

(262, a - d) In Kirsten Gynther Eriksen's workshop:
a. Kirsten Eriksen showing her molds.
b. King Henry VIII and supplies on work table.
c. Dolls in progress, and news articles about her dolls.
d. Artist with jumping jack; packaged dolls in background.

Ill. 263

Ill. 264

(263) Pamela Nedergaard's childhood teddy.
(264) Pamela Nedergaard with Kirsten Eriksen.

Kirsten Eriksen's historical porcelain head figures differ from the beautiful queen and other porcelain head dolls, highly glazed, made by the Royal Copenhagen firm in the 1840s. The distinction between the early porcelain or china, produced from 1840 into the early 20th century, and today's porcelain is created in the finishing. Old china heads are highly glazed. Today's bisque porcelain heads are unglazed with a matte surface; the material is often tinted before pouring, and the features are delicately colored, in contrast with the bright colored eyes, lips and hair of the glazed porcelain faces. Today's artist dolls, and reproductions in the U.S.A. and in Europe are unglazed and closely resemble the early bisque heads rather than the antique porcelain called china heads.

This talented artist's goal, she told us in 1989, is to perfect one new doll each year. As described in Chapter 2, her 1990 efforts have far exceeded that goal, and the challenge of creating dolls of living persons has been most rewarding. Her hobby has become a full time business; it takes at least a week to create a doll. The dolls are 15 3/4 - 20 3/4 in (40 - 50 cm) tall. They sell readily in foreign markets in San Francisco, Dallas, and Japan. In Denmark they are sold to collectors at the Euro Toy Show and Sale, a twice yearly event at Legoland Park, and are available in a shop on the *Radhuspladsen* in Copenhagen. Each doll of one design is numbered. Kirsten said she sold over 40 King Christian IV dolls during his 340 year jubilee.

This very busy lady devoted an entire morning to showing her dolls, her workshop and molds, her research and costuming rooms in their beautiful farm home. She graciously served Pamela Nedergaard and her daughter and us their delicious cherry brandy, a product of their orchards.

Pamela Nedergaard

Pamela Nedergaard had lost her favorite dolls when she was evacuated with other children to the countryside during the German aerial attacks upon England. But she and her Steiff teddy bear (Ill.263) survived. Their childhood losses inspired both Pamela and Kirsten Eriksen to artistic creativity in their adult years. The head of Pamela's first cloth doll made in 1979 was molded of Cernit, a plastic baked in a kitchen oven. The doll was costumed in the fabric of her daughter's dress and given to a friend's daughter on the occasion of her confirmation.

Pamela (Ill.264) has since worked in several media. She has made dolls representing folk costumed dolls shown in Chapter 3 and wondrous Nisser shown in Chapter 5. Her little doll made with a round wooden head became a folk doll of Odense, her current home.

In 1985 one of those first dolls was presented to Her Majesty Queen Margrethe when Royalty visited the Odense University.

Pamela began making needle sculpted dolls in 1988; a pair represented an elderly couple in folk dress. Her "Traveling Minstrels", an ensemble of performing mice, was displayed at Tivoli in 1988. The variety of her imaginative needle sculpted Nisser has been documented in Chapter 5.

Pamela had explained that her interest in needle sculpting "is perhaps a mixture of all things. I had seen soft sculpting in *Doll Reader* as a craft. I had read Catherin Christopher's (paper back) book on doll making, (pages 41 and 42), with her pictures of the doll "Helmi". It was this doll I saw in a shop window in south Germany used as a window prop. The shop was closed so I couldn't find out anything about her. It was a combination of impulses which got me started (on needle sculpting)" [86]. Using this medium she produces an endless variety of personalities of doll folk as regional costumed dolls and Nisser (See Ill.65, 123-125).

Pamela had graciously arranged our first visit with Kirsten Eriksen. We enjoyed the Nedergaard's hospitality in their Odense home and meeting her pharmacologist husband and younger daughter. Their personal doll collections have also provided documentation of many Danish celluloid dolls in Chapter 8.

Pamela has become involved in organizing a doll club, and is currently writing articles for doll magazines. Her contributions to Danish doll knowledge are most welcome.

Ill. *265*

(265) Aase Breivik with Salling dolls at Artists' reception.

Aase Brevik

Mrs. Aase Breivik, Norwegian by birth, has brought to Danish doll design a remarkable talent, as had the English born Pamela Nedergaard. Mrs. Breivik has been a textile designer for many years, and taught needlecrafts. In the fall of 1988 she learned doll making and began teaching classes in January of the next year. She found it "the right thing for me; I sew, paint, and design clothes. I draw my dolls as paper dolls for *folkedragter*." [87]. This craft, making dolls utilizing commercially made plastic face molds to be covered with knitted fabric and painted, drew upon all her talents and proved very exciting to teach, especially dressing the dolls in folk costumes.

She has made wigs for her dolls from knitting yarn or from lambskin. Some ready made wigs of synthetic materials are used.

Her doll making classes were extremely popular in her town of Gentofte and in Copenhagen, and are now in demand in Bergen, Norway, where she moved in January

Ill. 266

Ill. 267 a

Ill. 267 b

(266) Thomas Dam with wood carving of troll. ®Dam. Courtesy of Niels Dam.
(267 a) Dam Lucky Trolls of 1990.
(267 b) Dam trolls, Norfins Kids, articulated at shoulders and hips, introduced in 1992. Courtesy of Eva Stark, E.F.S. Mkt. Assoc. Inc.

1990. Much of her inspiration for authentic folk dress of Denmark has been the illustrations of costumes in the National Museum in Copenhagen, documented in Minna Kragelund's book which is referenced in Chapter 3.

Aase and her dolls dressed in Salling folk costumes were shown and photographed (Ill.265) at a reception at which we hosted doll artists from several places in Denmark, in October 1989. She expressed a desire to learn to mold the doll heads and thus make dolls which would be entirely original. Her dolls have been dressed in Norwegian folk costumes as well as Danish, incorporating her love of embroidery and authenticity in designing the doll costumes.

She had requests to continue teaching doll making in Norway before moving there. At crafts fairs in Bergen four of her Norwegian dressed folk dolls had been exhibited and applauded. Those in Danish folk dress are also widely exhibited.

Thomas Dam

The fantastic success of the Danish trolls created by Mr.Thomas Dam should make him the father of that symbol of Denmark today. His wood carving, hanging above the desk of his son Niels, was inspiration for the doll sized carvings (Ill.266) which ultimately resulted in the manufacture of the trolls. The involvement of all the family in today's production was told in Chapter 4.

Subtle changes in the designs of current trolls can be seen in the faces of little Lucky Trolls (Ill.267 a) brought to our grandchildren in April 1990. These are about 8 in (20 cm) tall. Each has a different colored wig of acrylic; six year old Shandra arranged many hair styles on her "Lucky Looie". And each troll wears a bright colored felt tunic, characteristic dress of the 1960s trolls.

The trolls were designed by the talented carver, Mr. Dam, who died on Nov.12, 1989. He has left an everlasting heritage in this line of dolls and toys. His happy looking trolls have changed the image of these wee people from the stern and serious characters of Scandinavian folklore to lovable, impish creatures. They become welcome additions to collectors of dolls and related toys. Finding the trolls described in doll books and doll magazines for over 15 years confirms their deserved recognition in the expanding realm of Danish doll related collectibles. The articulated Norfins are 1992 Dam designs, made doll-like with neck and shoulder articulation of the doctor, nurse, and school kids, and the fully jointed Norfin Kids (Ill.267 b). The Dam family's talents and successes verify this entry in the presentation of artists in Denmark.

Mrs. Ester Pedersen

Mrs. Ester Pedersen of Kastrup, south of Copenhagen, learned doll making in a class after a long and boring convalescence following her broken back. This 73 year young lady had made numerous cloth dolls of tricot faces over plastic molds. Seen in Chapter 4 were her pair of dolls photographed at the Israels Plad flea market and our Miss Blue Pedersen bought there. The body construction of Miss Blue was compared with that of the two Nisser she brought to our 1989 meeting of doll artists. Her choice of wigs and the styles in which she dresses them give her cloth dolls unique personalities.

Mrs. Pedersen brought several dolls which showed her skill in working with other materials . Bisque porcelain dolls were her three-faced black baby doll similar to Mr. Risom's doll shown in Chapter 7, a large bisque baby doll (Ill.268), and a *Julemand* made from an S.F.B.J. mold. A beautiful wax doll she is holding is another accomplishment. Mrs. Pedersen was challenged by wax doll making; she bought the first wax available within four hours after the craft supply store had received it. A restored doll of which she is justifiably proud is the carnival character (Ill.269) resulting from her salvage of a discarded broken bisque head.

The *Julemand* in his modern red jumpsuit, her Nisser and wee snowmen and Nisser made of knitted costumes covering styrofoam balls sparked a real holiday mood at our meeting of doll artists. Mrs. Pedersen's enthusiasm was endearing to all of us.

Ill. 268

Ill. 269

Jette Vibe

Occasionally the creation of a new doll or toy is the result of the artist's experience in making up commercial designs or kits. Mrs. Jette Viby of Holte began doll making in 1960, completing purchased kits for their four daughters. The year 1970 began what she called her second creative period, making her own original designs. Her cloth dolls, animals, games, and wall hangings have been enthusiastically received and are still popular. Her Millie doll shown in Chapter 6 was copyrighted in 1979; it continues to sell well. Current efforts are primarily wall hangings. The newest project is a secret, she told me in a May 29, 1989 interview. Mr. Andersen in Toftlund, Jutland now manufactures her designs, which are sold in kits, printed cloth sheets with directions in Danish and in English.

Ill. 270

(268) Ester Pedersen with two of her bisque dolls and wax doll.
(269) Ester Pedersen's salvaged carnival doll.
(270) Jette Vibe with her cloth dolls and stuffed animals.
© courtesy of Jette Vibe.

Den Permanente, the store which for many years offered the choice of Danish designed items, sold Mrs. Viby's kits. Our Millie kit bought there was our introduction to her creativity. Recent or current markets have included *Illums Bolighus*, and *Haandarbeijdets*, an exclusive needlework shop on the Stroget in Copenhagen, Neiman-Marcus in the U.S.A., and Harrod's in London. A shop in Japan also sells her creations.

Prestigious recognition has been international. For "The Year of the Child" emphasis at the United Nations in New York City in 1979, Mrs. Viby's Noah's Ark with many stuffed animals was one of only two Danish designs chosen for exhibition. The B.B.C. in London has used her dog and monkey in a serialized television show, and the Japanese Tokaido Museum has exhibited her work.

In October 1989 Mrs. Vibe shared with us pictures of some of the vast output of 19 years. She has graciously given permission to use a photograph of herself with some of her classic animals and dolls (Ill.270).

Camme Bache

A telephone call to Praestø and letters with questions have brought another artisan into this book. The twelve well made doll house dolls by Camme Bache are described in Chapter 11.

In 1977 Camme Bache had just opened her museum, *Den lille By i Praestø*, (the little town in Praestø). A friend advised her to send for the books by Mildred Seeley, an American authority on making reproduction dolls, now living in Green Valley, AZ. She went to Rome where Mrs. Seeley was at that time. "She told me everything she knew about doll making, and she knew a lot" [88]. Back in Praestø she found no one who knew about doll making; she found no Danish firm which sold the needed materials, so those all had to be imported. She found a Dutch made kiln which had no automatic controls and needed constant attention for accurate firing. Many problems had to be solved to even begin, and her beautiful doll house dolls required much effort to perfect. Eventually instant success was experienced in sales at the gift shop at Legoland Park where we bought the doll house man doll, and at other outlets.

Camme Bache's museum was closed in 1991, with an auction by Christie's of London. Many of the exhibits are relocated to the *Puppenhaus Toshino Museum Collection* in Tokyo, opening in April 1992.

Astrid Bernth

Astrid Bernth, Swiss born and educated, now lives in Silkeborg, Jutland. Her stick puppets (Ill. 271) and delightful character and caricature dolls are one-of-a-kind, which she molds of her own formulated clay.

"I make many types of dolls. There are dolls for play, dolls for decoration, theatrical dolls, marionettes and stick puppets. There are soft dolls, hard dolls, ugly dolls, fearful dolls, and dolls for hugging", she wrote. The originality of those dolls cannot be questioned; those shown in Chapter 10 are unforgettable.

Teaching students and adults remains a very important part of her doll creativity and generosity. She inspires artisans who use commercial molds to try molding their own doll heads.

John Nissen and Else Madsen

A sought after designer and sculptor in Copenhagen, John Nissen, and the co-owner and founder of Dan-Hill-Plast a/s in Hornsyld, Mrs. Else Madsen, very successfully produce exquisite fashion dolls and dolls of international design.

In 1947 John Nissen began his distinguished career as a sculptor at the Royal Danish Porcelain Company in Copenhagen. In 1960 he became chief designer for *Hindsgaul*, the largest manikin manufacturer in the world, and in 1973 he started his own company.

Visiting *Hindsgaul* we wandered through a large room filled with life sized manikins, so realistic that we expected animated conversation. We were shown his models for medical training, one to be used for resuscitation of drowning victims. In his office (Ill.272) we were treated to a viewing of the completed and wigged heads of the newest series. "United Children of Our World". He showed us the limbs made for these well researched dolls, scheduled to be introduced at the 1990 Nuremberg Toy Fair. In October 1990 doll advertisements the Dynasty Doll Collection in New Jersey presented these dolls, now available in doll shops.

Many of the dolls documented in Chapter 9, the series called "The World's Children", and the Nigerian family were Mr. Nissen's creations. The manikins or fashion dolls representing the decades from 1910 through 1980 were designed by him. Their unique construction detail, the removable arms, facilitates dressing in those numerous styles. I found this to be true, undressing and examining the manikin doll Jessie.

The very accurate and beautiful costumes worn by those manikin dolls and the international dolls are designs resulting from Mrs. Else Madsen's conscientious research. Each detail is carefully scrutinized before an ensemble can be labeled with her signature. This meticulous attention to design and construction is also very apparent in her costumes of the other well made dolls, which are sold to kindergartens and play schools. All this care and cncern is reflected in the guarantees with which the dolls are sold.

Graciously given was the time spent with us on our visits to both the Hornsyld firm and *Hindsgaul.* Mr. and Mrs. Madsen and Mr. Paul Hansen at Dan-Hill-Plast showed dolls and toys, and answered all our questions. Mrs. Madsen translated archives and toured the factory with us, and thrilled me with the gift of the 1920 doll Jessie (Ill.273). Continued correspondence has provided the photographs and documentation of their newest series, "United Children of Our World", added to their story in Chapter 9. The Madsens have sent the Northern European doll head of that series to model an exquisite black silk Amager cap at the Danish Immigrant Museum in Elk Horn, Iowa.

This artist-sculptor and designer pair, John Nissen and Else Madsen, will continue to delight the doll world with their unique and accurately designed creations.

This presentation of the doll artists and artisans whose work has been described and pictured in the preceding chapters has shown that doll making today in Denmark is extensive and diversified, a fascinating study. Any one of these artist dolls would be a delight to the collector of contemporary, folk or historical dolls.

Ill. 271

Ill. 272

Ill. 273

(271) Astrid Bernth with her stick puppet.
(272) Author and John Nissen in his office at *Hindsgaul.*
(273) Else Madsen presenting Jesse doll to author.

Chapter 13

Museums in Denmark Exhibiting Dolls

Where can the collector or historian look for dolls in Denmark? They can be seen in museums from the island of Fanø off the west coast to Copenhagen and Køge on the east. Some museums were established exclusively for dolls, doll houses, and related toys. They are also found in collections in national museums. Rare English, German and French dolls described command the attention of visitors. An excellent representation of documented antique dolls is among the numerous attractions at LEGOLAND Park® at Billund, the second largest tourist attraction in Denmark. This vast enterprise of 21 amusements and activities opened on June 7, 1968. In this chapter describing museum collections only three areas are presented, some of the dolls, "Titiana's Palace", and the display of the shipwrecked bark *Louisana*.

LEGOLAND®
The Doll Collection

In 1966 LEGOLAND A/S acquired the lifelong collection of Mrs. Estrid Faurhold, begun, she explained, when she inherited her grandmother's 1870 dollhouse on a memorable childhood Christmas [89]. The collection of about 500 dolls and doll houses was vastly increased by the 1989 acquisition from a Copenhagen collector. The doll houses and dolls span more than three centuries. "Madeleine Reynaud" (Ill.274), the oldest dated 1580, is a Gothic appearing doll carved from wood. A rare 1690 wooden doll, "Lady Elton", and the brown leather pair, "Victoria" and "Albert", ca 1850 - 1860", are English (Ill.275). An elaborate French fashion doll is Odile, 1875, exquisitely gowned in ecru silk and rust velvet (Ill.276). Dolls and doll heads represent decades of papier mâché, porcelain (china), parian type and bisque beauties. Others we studied in 1979 include mechanical dolls, shell dolls dated 1840, Japanese babies, and doll house dolls. In 1989 we saw additions of the early 1900s,

Ill. 274

Ill. 276

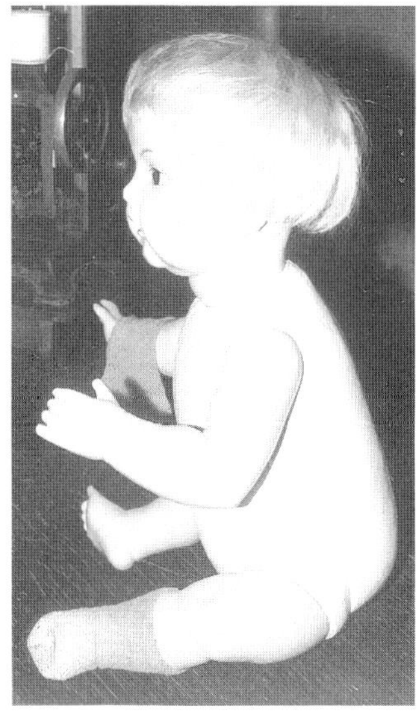

Ill. 277

Ill. 275

Ill. 278

(274) "Madeleine Reynaud", 1580 carved wooden doll. © LEGOLAND, courtesy of Kirsten Martin Stadelhofer.

(275) "Victoria" and "Albert", English, 1850 - 1860. © LEGOLAND, courtesy of Kirsten Martin Stadelhofer.

(276) "Odile", French fashion doll, 1875. © LEGOLAND, courtesy of Kirsten Martin Stadelhofer.

(277) Schildkröt doll with unusual body. Photographed courtesy of Kirsten Martin Stadelhofer.

(278) Simple home made doll house. Photographed courtesy of Kirsten Martin Stadelhofer.

favorite German bisque character children, and a celluloid head Schildkröt doll with an unusual body (Ill.277). Doll houses, from plain home crafted models (Ill.278) to elegant mansions, and the exquisite "Titiana's Palace" (Ill.279) present an unbelievable realm of miniature excellence.

Ill. 279

Ill. 280

(279) Titiana's Palace, © LEGOLAND, courtesy of Kirsten Martin Stadelhofer.

(280) The Royal Bedchamber. © LEGOLAND, courtesy of Kirsten Martin Stadelhofer.

LEGOLAND A/S
"Titiana's Palace"®

Thirty-two pages in Christie's Auction catalogue of 1978 were devoted to listing the fantastic work of art known as "Titania's Palace" [90]. Purchased by LEGOLAND A/S , this "Fairy Queen's Miniature Palace®" became a star attraction when introduced in 1980.

Titiana, the Queen of the fairies in Shakespeare's play, *A Midsummer Night's Dream*, appears throughout her palace as a white butterfly with orange tipped wings. Sir Neville Wilkinson, a retired officer from the Boer War, planned this elaborate project at the request of his little daughter Gwendolyn. Begun in a Dublin carpenter's workshop in 1907, the palace was opened by Queen Mary of England in 1922. Sir Neville added untiringly to the furnishing of the 18 halls and rooms. Today over 3,000 tiny pieces, many copies of fine furniture, antiques and art work, all the castle's opulent furnishings, have been photographed and catalogued individually. The measurements of the building, on the scale of 1:12, are 92 X 118 in, with the bell tower reaching the height of 74 in (230 X 295 X 185 cm).

In the Hall of the Fairy Kiss the highest order of Fairyland is kept in a crown-topped chest. Above the chest is the Fairy Queen's motto in Latin,

"NIHIL SINE LABORE"
("Nothing without work").

The Fairy Family receives private guests in an Entrance Hall with double staircases. There the Visitors' Book is placed on an opal case topped with a crown. Small statues of silver and mosaic windows decorate the stair setting. The Chinese silver filigree lamps were donated by Queen Mary, herself a collector of doll house miniatures. Rare furnishings and accessories decorate other rooms in the palace, particularly the four poster bed with ivory carvings, the red lacquered chairs, and gold and ivory jewelry boxes in the Royal Bedchamber (Ill.280). Miniature religious items in the Chapel include a New Testament and other books, a copy of a Madonna by Murillo, a rosary said to be the smallest in the world, and an organ which can be played. The Royal Dining Room has an elegantly set table and a sideboard made of satin wood. Details of the fine furnishings are seen in the red lacquered secretary (Ill.281) in the Morning Room. The secretary is only 3 in (7.6 cm) wide at the desk top. That is less than twice the width of the wingspan, 1 3/4 in (4.5 cm) of the butterfly, shown life size with many actual size items in the castle throughout the little book, *TITIANA'S PALACE* [91].

LEGOLAND A/S
Shipwrecked Dolls

In January 1868 the 1,300 ton bark *Louisana* left the Bremerhaven, Germany shipyard on its maiden voyage with cargo and a crew of 22. Attempting to reach the English Channel, the ship encountered a violent storm; mast and rigging were washed overboard, and the helpless ship drifted across the North Sea for two days. When the sinking vessel was seen from the shore of Denmark near Tuskver, a lifeboat was dispatched to help. One seaman drowned; the others were rescued. Salvaging operations were begun, but too quickly the ship sank into the sandy sea bottom and was pounded to pieces.

More than a century passed before the sea was to surrender the rest of the cargo. In 1983 divers found wreckage at seven to eight meters, 500 meters off the shore of Fjaltring. In the bottom of the hold were many undisturbed boxes in which were found ten doll heads, and several pieces of china tea sets.

"The most exciting part of the cargo was the doll parts. Heads, arms and legs, plus small cups, plates, dishes and jugs were found all over the wreck." [92].

Today's LEGOLAND A/S museum visitor finds in a circular hall the story of the *Louisana* documented and illustrated with artifacts in realistic displays. Some doll heads with seaweed and seashells can be studied closely within small containers. One salvaged doll head is still well detailed with little apparent damage. The recreated sea bed displays spoils of the wreckage strewn over the ocean floor. Kirsten Martin Stadelhofer posed with some of the findings (Ill.282).

In order to take in all the wonders of the doll exhibitions and any of the other 18 attractions one should spend more than a day there. We know that we shall learn more on our fourth visit.

Ill. 281

Ill. 282

(281) Red lacquered secretary. © LEGOLAND, courtesy of Kirsten Martin Stadelhofer.

(282) Recreated sea floor, *Louisana* wreck, with Kirsten Martin Stadelhofer.

Ill. 283

Ill. 284

Ill. 285

(283) Käthe Kruse dolls, "Peter" and "Emil'. © Courtesy of Doris Poulsen.

(284) German bisque boys in typical costumes. © Courtesy of Doris Poulsen.

(285) German bisque head dolls, each 23 5/8 in., "Ida" by Kling, and "Dora" by Schoenau and Hoffmeister. © Courtesy of Doris Poulsen.

LEGETOJMUSEUM, Doris Poulsen's Toy Museum at Randers

A fine private museum was opened by Doris Poulsen in 1981, after she acquired "a big and beautiful collection which I thought would interest others." [93]. She found her first antique doll in 1965 and couldn't resist acquiring other dolls. Of special interest are the Käthe Kruse boys (Ill.283) "Peter" and "Emil", named for readily recognized runaway characters in a children's book. Peter is dated 1912; Emil is a later 1938 doll. Another pair of 1912 bisque head boys with molded hair, "Kristian", almost 16 in (40 cm), and 13 in (33 cm) "Peter" (284), wear handsome suits with below-the-knee pants. Fashionably dressed are some large bisque head dolls quite rare in private collections in the United States. C. F. Kling dolls of 1885 are 25 1/2 in (65 cm) "Harriet" in white eyelet, 23 5/8 in (60 cm) "Ida" (with "Dora", Ill.285), and 13 3/4 in (35 cm) "Elizabeth". Two brown eyed bisque head dolls are the A.M. "Margrethe", 13 3/4 in (35 cm), and Schoenau and Hoffmeister's "Dora", 23 5/8 in (60 cm), dated ca 1909. A baby doll with bisque head and the boy dolls represent the character dolls popular after 1910. The rare black Sara and Tom were shown in Ch.7.

The museum, located at Stemannsgade 17, 8900 Randers, has added a small boutique for visitors wanting souvenirs.

Ill. 286

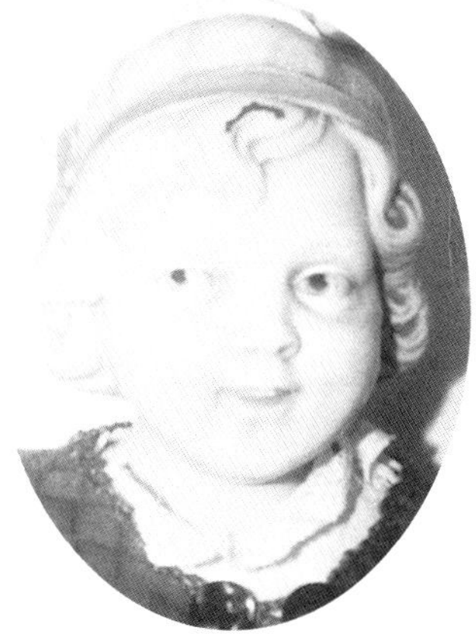

Ill. 287

(286) Dolls in MISSERs Doll Museum, Ebeltoft.
(287) Celluloid head with a Heubach look, Ebeltoft.

DEN GAMLE BY, (The Old Town) at Aarhus

This fabulous open air museum, with over 50 houses and buildings moved here from a dozen Danish towns and refurnished, offers a first hand look at Danish living conditons from 1600. Two buildings on Sondergade contain displays of interest to doll collectors. Number 3a, *Textilsamlingen* (textile collection) contains toys and dolls. Many are beautiful porcelain dolls; none could be verified as Danish dolls of the 1840s. One which resembles the Royal Copenhagen porcelain dolls could be the German K.P.M. Another tall china head doll, ca 1865, has black hair, a body of leather, and legs and arms of wood. Several dolls in old carriages, and wonderful doll houses are exhibited, one dated 1855. In building 7 across the street is the glover's shop where leather gloves and bodies for dolls were made. Displayed with chamois leather doll bodies are lovely old porcelain and bisque doll heads. Permission to use our photographs of the dolls and doll houses described was not allowed. The collection at Old Town "contains about 700 dolls, so only a small part (but the best one) is exhibited. The oldest one dates from about 1810, and from the 1840s, we have every decade very well represented up to (the) present time." [94]. This outdoor museum can be contacted at DK-8000, Aarhus in Jutland.

MISSER's DUKKEMUSEUM; Doll Museum at Ebeltoft

In the bayside village of Ebeltoft, 46 kilometers from the Aarhus airport, a small stucco cottage houses many dolls in two rooms, on the floor, walls, and shelves (Ill.286). There are German bisque head dolls and Danish and German celluloids. Delightful surprises were a celluloid doll resembling Heubach's bisque Coquette with open-closed mouth (Ill.287), the Greenland folk doll with the very old Nisse in Chapter 5, and a framed panel of *Knold and Tot*, the *spraellemaend* shown in Chapter 11. A doll house, china dishes, and other accessories added interest. In the absence of labels one could only guess the identity of many dolls not readily recognized.

INTERNATIONALT HISTORISK LEGETOJMUSEUM at Nordby, Fanø

A privately owned museum is located in another quaint cottage at Herluf Trollesvej 1, Nordby, 6720 Fanø. The owner is Mr. Ole Jens Bruun, a civil engineer in Assens, Funin. A museum poster shows an elaborate doll-house and dolls; another shows many Danish toy soldiers. We have not seen the museum and have no details about the dolls there, except several Skookum Bully Good Indian dolls we took to Denmark in 1989.

KØGE MUSEUM

In a 17th century merchant's house at Norregade #4 in Køge the history of civilization of that area is exhibited. The museum has occupied the house since 1910. Toys and dolls are displayed in room 17 under the eaves on the third floor (called the second floor in Europe). Not many of the 66 dolls are exhibited at one time. In July 1990 those were German china head dolls and one with a bisque head, an old home made doll, and some doll dishes (Ill.288 a, b). Marie Bach, Museum Inspector, wrote that five dolls there had been made in the Køge factories of J. K. Kofod. The dolls and toys are a very small but interesting part of this Danish museum.

DUKKEMUSEET: Mme. Margot Siesbye's Antique Doll Museum

A doll collection of two or more decades has become an exciting private museum at 6, Kronprinsessegade in Copenhagen, open all year by telephone appointment. Margot Siesbye's dolls are housed in upstairs quarters above former stables facing the court behind her home. The collection, begun with one antique doll the Siesbyes had bought years ago, has grown to include over 1,000 dolls and miniatures. During our timely visit we photographed Mme. Siesbye with her newest acquisition, the girl doll by Royal Copenhagen (Ill.289) and googlies (Ill.290). A variety of china and bisque head dolls from Germany include beautiful dolly face and character dolls. All are displayed, uncrowded, some with historic scenes as backgrounds. Doll houses and miniatures are in another wing. A photograph of Her Majesty Queen Ingrid at the opening ceremony graces one of the walls.

Mme. Siesbye has received dolls as gifts for her museum. She contributes the entry fees to the Danish Red Cross. Our doll bed collection is enhanced by the mahogany bed, ca 1880, bought there (Ill.291.). Mme. Siesbye's gracious informality makes visiting this collection a delight for anyone, and a treasure hunt for serious dollers visiting Copenhagen.

Ill. 288 a

Ill. 288 b

(288, a, b) Dolls and dishes displayed at Køge museum in 1990. Photos courtesy of a friend.

(289) Mme. Siesbye with latest doll, a Royal Copenhagen porcelain, in her *Dukkemuseet*, Copenhagen.

Ill. 289

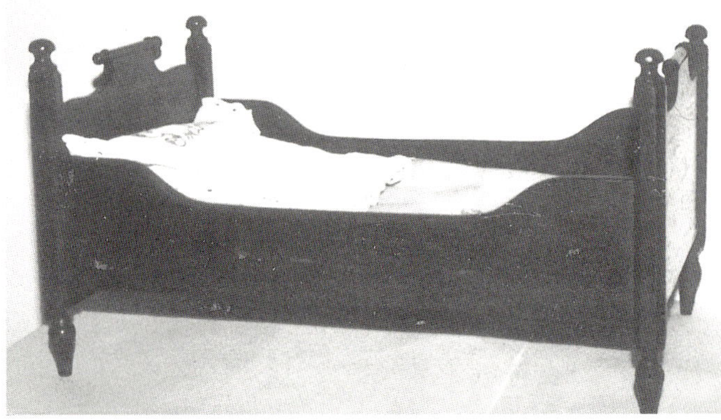

Ill. 291

Ill. 292

(290) Googlies in *Dukkemuseet*, Copenhagen.
(291) Mahogany doll bed, ca 1880, from *Dukkemuseet*.
(292) German china head doll, ca 1840, in *Legetojmuseet* in Copenhagen.

LEGETOJMUSEET: The Toy Museum in Copenhagen

In 1979 we had found a small museum, very crowded but very exciting in its offering of dolls and toys. This Toy Museum now occupies handsome, spacious quarters at 13, Valkendorfsgade, just a brisk walk from the *Stroget*. Mr. Fritz Hartz began the museum with his own extensive collection, and other collectors have donated many items. Antique dolls and Victorian dollhouses, doll furniture, and celluloid dolls in settings, and a row of doll carriages are appealing. A tall, exquisite china head doll, ca 1840, was brought from Germany to Denmark by its owner, a 14 year old bride (Ill.292). A pair of the famous hedgehogs, Mecki and Mucki, the largest we have ever seen, and other old Steiff animals having tea (Ill.293 a, b) are some of many delights for Steiff lovers. Over 2,500 items include steam driven and wind up toys and magic lantern shows. "The museum is no graveyard of old toys - it is a *living* museum with entertainment for children including a dinky kiddies' cinema - Denmark's smallest with only 30 seats!" [95]. Children and adults shouldn't miss it.

Ill. 293 a

Ill. 293 b

(293 a, b) Steiff animals at tea., in *Legetojmuseet*.

Chapter 14

Dukken Margrete in the Danish National Museum

"K*jaere Maggie,*

Vil Du sporge Din Moder om Du maa komme til mig i morgen Kl. 12, hvis har Lyst. Vi Skulle lege med Dukkerne, saa du vil maaske tage din Dukke med. i Haab om at see dig forbliver jeg...(ulaeselige ord) hengivne.

Ven eders,
Margareta Wolff" [96]

"Dear Maggie,

Will you please ask your mother if you may visit me tomorrow at 12 o'clock? We will play with our dolls if you will perhaps bring yours. Hoping to see you I remain ... (illegible words) devoted friend.

Yours,
Margareta Wolff"

That is how a seven or eight year old girl wrote from her doll to herself, to Maggie, the nickname of Margarethe, or Margarete as she signed her name. The doll was named Margrete Wolff, after its doll mother.

Maggie's father bought the beautiful 20 1/8 in (51 cm) wax doll in Paris for her June 1, 1860 birthday. This English import was probably made by the Montanari family, who had displayed their dolls at the Great Exhibition in 1851 and at the Paris Exposition in 1855 [97]. The dolls were "characterized by their plump, rather short necks with a roll of fat in the front" [98]. Three German experts on wax dolls, R. and H. Knedel of *Das Frankfurter PuppenHaus*, Frankfurt am Main, and G. Gauder [99] of Stuttgart, on viewing transparencies of the doll Margrete, gave further support to the suggested identity. The color of the wax is much lighter than the reddish color of the wax dolls made by Pierotti at that time. And the spot on the chest of the cloth body (Ill.294) is where the Montanari label would have been glued.

The letter which introduces the story of the doll Margrete is a folded paper, 3/4 in X 1 1/8 in (1.7 X 2.8 cm) addressed

> "to Miss M. Wolff, Little *Fiolstraede* (Fiddle Street) No.10, Copenhagen, in front of the University" [100].

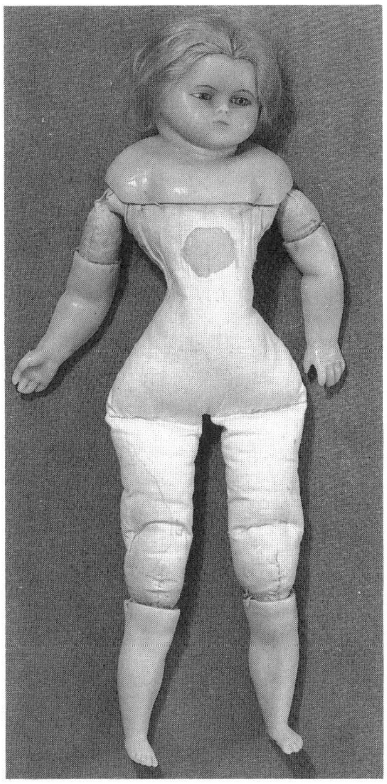

Ill. 294

Ill. 296

Ill. 295

Ill. 297

Ill. 298

(294) The doll Margrete, undressed. *Nationalmuseet*, Niels Elswing photo.

(295) Photograph of Maggie Wolff in decollete dress. *Nationalmuseet*, Niels Elswing photo.

(296) Doll wearing dress styled after Maggie's in Ill.295. Eileen van Baren sketch after N. Elswing photo.

(297) Doll wearing the blue dress and bonnet.

(298) Doll's 2-piece dress of same blue fabric as dress in Ill. 297. Photographed at *Nationalmuseet*.

Ill. 299

(299) Floor length dress of printed mull. Photographed at *Nationalmuseet.*

Maggie lived there with her two brothers, one older and one younger, and her mother and father. Mr. Ferdinand Wolff, a wealthy banker, had been a merchant and then the Danish Consul to the Philippines. He had traveled widely and collected extensively, as was reflected in their home and their dress.

The daughters of wealthy Danish families never attended school. Maggie, with two friends about her age, were tutored at her home. The tiny accessories in the doll's backpack included grade books with daily grades in these subjects: religion, English, French, Danish, natural history, drawing, writing, knitting, sewing, arithmetic, geography, history, and spelling. These courses revealed the importance of languages in Maggie's home. Eight little blue covered composition books, 1 5/8 in X 1 15/16 in (4 X 5 cm), were for deportment, drawing, and French dictation. The grades given to M. Wolff in sewing and knitting were very high, and were probably a reflection of Maggie's interest in those domestic arts. Maggie may have made many of the doll's clothes, and may have been assisted by Sophie Frels, the godmother of her younger brother and a close family friend. Contrary to common practice, there was no known live-in seamstress employed by the Wolff family.

Both Maggie and her mother were considered very well dressed at all times. The doll Margrete had a wardrobe of costumes of the finest materials such as Oriental silks and other unusual imported fabrics. Her garments and accessories, numbering over 160, give today's reader a picture of Maggie's life in the 1860s.

The Danish National Museum has catalogued the following [101]:

Chemises	2	Hats & bonnets	9
Pantaloons	2	Mittens & wristlets	4
Underskirts	7	Jewelry	3
Nightgown	1	Handkerchiefs & case	19
Bodices	8	Pillow slips	3
Dresses	10	Bags & purses	6
Aprons	3	School bag & contents	18
Bibs	4	Baskets	9
Shoes & stockings	17	Boxes	3
Jackets	5	Dinner service	5
Outdoor things	4	Sundries	12

The doll's ten dresses include both one-piece and two-piece styles, often in the pattern of those worn by Maggie and her friends. One lovely dress made in a style and fabric similar to that of the decollete dress worn by Maggie (Ill.295) is shown in a sketch made after the doll's photograph (Ill.296). A rare, loosely woven fabric, horizontally striped with blue wool and white cotton, was used for this two-piece daytime dress. The midcalf length skirt, 10 1/4 in (26 cm), is unusually full, measuring 45 11/16 in (116 cm) around its narrow hem. The waistband of the skirt measures 9 13/16 in (25 cm), and laps a bit more than 3/8 in (1 cm) to fit the doll's 9 7/16 in (24 cm) waist. The bodice is 4 3/4 in (12 cm) from the neckline to the waist, including a 5/8 in (1.5 cm) tuck-in. The gathered waistline is stabilized on a ribbon stay. The neckline ruffle, about 3/8 in (1 cm) wide, is created by a string through a casing which ties at the back. More strings through casings gather the puff sleeves. The dress and a matching bonnet were displayed on Margrete during our photographing session at the museum (Ill.297). The neckline, drawn up tightly, and the wide blue ribbon bow tying the bonnet do not reveal the decollete style of this dress.

A second two-piece dress of the blue striped material (Ill.298) has a longer skirt, 13 in (33 cm), and a fitted, lined waist with four stays between the lining and the dress fabric. Blue silk ribbon binds the neck and ties at the center back in halter fashion. This may be considered the most formal of any of the doll's dresses.

Another elegant dress of printed sheer mull fabric, perhaps from Persia, is made with a very full floor length skirt, gathered into a wide waistband (Ill.299). The neckline is gathered into a narrow self-binding, and binding of the very full sleeves trims the armseye. Buttons on the cuffs of the sleeves, waistband, and placket are 1/8 in (3 cm) in diameter, sewn on with white thread, and fastened with either worked buttonholes or dainty thread loops. This dress was worn by the doll in the cover photo on the book, and also in the museum display case the last time we saw Margrete.

A dress of print on window pane checks (Ill.300) is one-piece with its very full bodice and skirt gathered into a snug waistband. The fully gathered sleeves can be worn off the shoulders. A double puff in the sleeves is created by a ribbon being pulled up from the lower edge of the sleeve to the top of the shoulder and buttoned. This interesting detail is shown in a sketch in Maggie's French story book, *Mlle. Lili.*

All of the doll's dresses are of light fabrics, in contrast with the heavy brocades and silks used extensively throughout Europe at that time. Two dresses are made of "reverse prints", also called "twin prints" or "positive and negative prints". These materials, in which the colored design on a white background becomes the white design on the colored background of the second print, are in vogue today.

When Maggie and her friends took Margrete for a walk, she probably put a short jacket or a cloak, a bell shaped garment, over the doll's light dress. A waist length cloak of birdseye patterned fabric fastens with one hook and eye under violet silk ribbon ties. A hat or bonnet was always worn. One white sun bonnet is made of the birdseye patterned cotton. Another bonnet of white lace has violet ribbon ties; either of those two would match the short white cloak. The fashionable "dame hat" is a small round of brown velvet trimmed with black lace and gold colored metal flowers. This stylish but totally impractical

Ill. 300

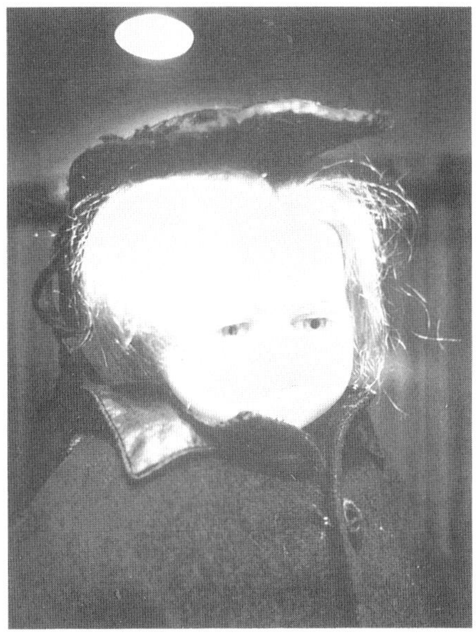

Ill. 301

(300) Dress of windowpane checks, showing unusual sleeve treatment. *Nationalmuseet.* Niels Elswing.

(301) Margrete wearing black coat and "dame hat". Photographed at *Nationalmuseet.*

Ill. 302

(302) Coat, and enlarged details. Eileen van Baren sketch.

hat complements a simply cut brown velvet jacket which is fastened at the neckline with one black button. This hat can also be worn with the long black coat (now aged to a brownish shade) in A line style (Ill.301). The coat is trimmed with black silk taffeta collar, cuffs, and pockets, all of which are edged with self cording. Three buttons at the coat front, two on each cuff and one on each pocket and on the points of the collar are covered with black silk thread, *passemont*, as were the buttons on the coat of the Vicar, Mrs. Nielsen's father. These details are sketched in Ill.302.

The doll probably would have carried a purse when visiting with her doll mother Maggie. Three of those six listed can be described. One is made of blue wool lined with silk; another is of black oilcloth, of which other accessories for the doll were made. A third, a little ecru lace edged pouch, is suspended from a blue ribbon belt which has a row of embroidery along the center.

A parasol was a necessity when ladies and girls took walks; fashion dictated a fair skin because a suntan would have indicated hard outside work. Margrete's *knaekparasol* is named for the clicking sound, *knaek*, made when the handle is folded in half. It is blue and beige silk edged with long beige silk fringe (Ill.303). The ivory trimmed handle is tipped with a blue glass ball.

Another accessory which could be used as flirtatiously as a fancy parasol was the fan. Margrete's was made of horn with studs of metal in the 1 1/8 in (3 cm) blades.

The undergarments made for the doll were fashionable as well as practical. One chemise, an undershirt, has short sleeves and a wide neckline to accomodate those decollete dresses. And another vest of quilted cotton material ties at the back with three ribbons. This was one of two garments needed for extra warmth, since the homes lacked central heating and were drafty. The other garment of wool (Ill.304, a) is a *sjaelevarmer*, a "soul warmer". It was made in a special technique called "Tunisian" crochet, of purple, black and white wool yarn. The unusual shape covers the doll's back to the waist, with two extended shoulder pieces to be brought down to the extended side pieces wrapped to the center front. The sketch (Ill.304,b) shows, better than words can describe, the way it was worn.

Two pairs of *mammelukker*, pantaloons, (Ill.305) are made of white cotton, both trimmed at the bottom of the legs with lace, for fashion's sake. In the mid-1860s crocheted white cotton trim on underwear had become quite fashionable, and probably the lace on Maggie's and Margrete's pantaloons was of crocheted cotton. One pair made for the doll has a drawstring to be tied at the back waist, and the center back opening continues up the center front for 8 in (20 cm). The center front seam is only 4 1/4 in (10.5 cm). The other pair is simply two wide legs to be worn on a drawstring around the waist. They needed heavy starching to be kept crisp, since their only function was for showing below the hem of a midcalf dress. These were called *kakkelovnsrorene*, "stove pipes", because of their straight leg shape.

The seven underskirts include two (Ill.306) for warmth, one knitted of white wool and another of red woolen fabric worn to help prevent "rheum". Either one would be worn beneath any one of the several crinolines designed to give the same effect as the hoop skirt, 9 1/2 in (24.2 cm) long and 8 in (20.3 cm) around, under a very full dress skirt. The doll's various skirt lengths, 10 1/4 in, 13 in, or 14 in, demanded crinolines of similar lengths.

The doll's cotton dimity nightgown is simply cut and constructed. The front and back are squares gathered into the neckline, which is finished with a collar edged in white lace. The sleeves are gathered into the armseyes and into the cuffs, also lace edged. Seven dimity covered buttons close the front, and each cuff is fastened with one button.

The doll's stockings and other hand knitted accessories demonstrate Maggie's accomplishments. Because knitting was considered an elegant pastime for young ladies and women, it was taught at a very early age. Margrete's stockings include a full length pair and a short pair made in stripes (Ill.307). Two pairs hand knitted of ribbed white cotton and two pairs of flesh colored wool yarn had been embroidered in grey cross stitch at the tops with M and W. A pair of footed stockings, styled like today's *après ski* sox, were made in stripes of heavy red and white wool. Other stockings were of machine knitted material. Rose silk garters were trimmed with tailored bows of matching silk ribbon.

White wool mittens were knitted with ribbed cuffs, with eyelets at the wrists for ribbon drawstrings; they are shown with the stockings in Ill.307. They are only about 2 in (5 cm) long, with very small thumbs. Knitted wristlets, just 1 3/16 in (2 cm), were made entirely in rib stitch, knit 2, purl 2. The doll had three pairs, also worn indoors for warmth and for relief from the discomfort of rheumatism, very prevalent in Denmark.

The museum inventory lists 17 pairs of footwear, of which 12 pairs were her stockings. Only one pair of Margrete's shoes is described in detail. They are black boots lined with dark blue silk, fastened with metal buckles and wide dark blue ribbon ties. The doll wears them with the blue dress and bonnet in which we photographed her at the museum.

The use of monograms became popular during the 1800s, and especially among the middle classes which were growing in number and in importance. Initials indicated the shift of status and became a sign of prestige as well as a decoration. Everything in a young woman's trousseau had to be monogrammed. Margrete's monogram, M, is angled across the corners of 13 pocket handkerchiefs. Eleven made of fine cotton batiste, 2 3/8 in (6 cm) square, have white monograms in chainstitch, and two are embroidered in red chainstitch. Two handkerchiefs, 2 3/4 in (7 cm) square, are fine white linen

Ill. 303

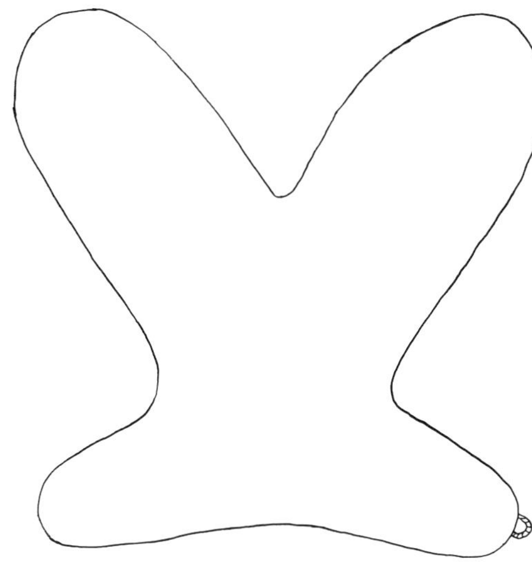

Ill. 304 a

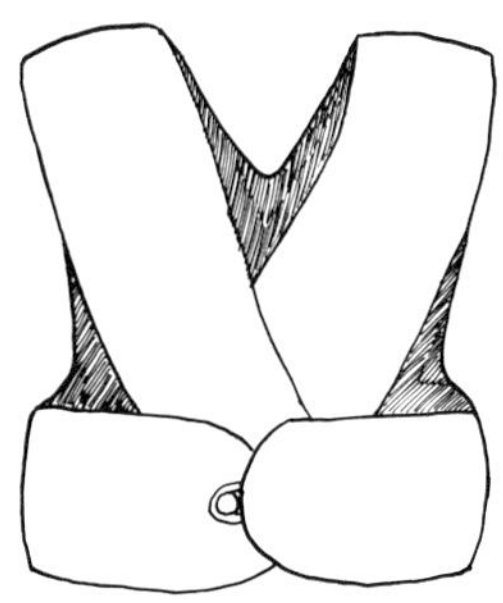

Ill. 304 b

(303) Parasol. Eileen van Baren sketch after Niels Elswing photo.

(304 a, b) "Soul warmer".
a. Half size, open.
b. As wrapped on doll. Eileen van Baren sketch.

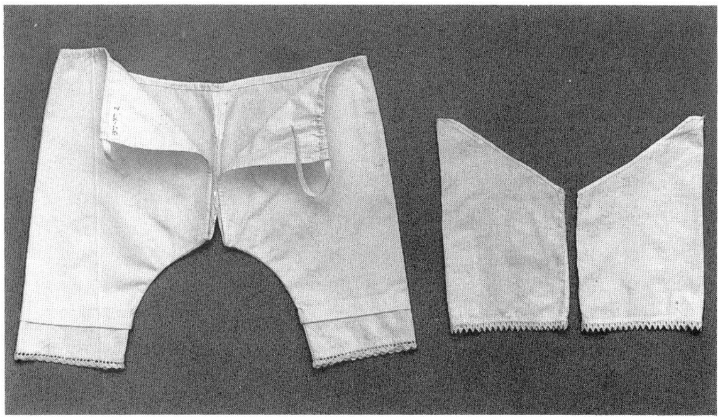

Ill. 305

Ill. 306 a, b

Ill. 307

(305) Pantaloons:
a. *Mammelukker* with partial center seam.
b. *Kakkelovnsrorene* (stovepipes). *Nationalmuseet.* Niels Elswing photo.
(306 a, b) Two warm underskirts.
a. Red wool.
b. White wool. *Nationalmuseet.* Niels Elswing photo.
(307) Two pairs hand knitted stockings and white mittens. *Nationalmuseet.* Niels Elswing photo.

embroidered in black silk featherstitch next to the neatly sewn hems. They are thought to be mourning handkerchiefs, although no deaths had occurred among Maggie's relatives in the 1860s. Had Maggie been inspired by the national mourning for King Frederick VII who had died on Nov.15, 1863?

Margrete's 18 handkerchiefs are neatly kept in a very elegant handkerchief case of padded pink silk edged with lace. Violet silk ribbon, 3/4 in (2 cm) wide, was used for hinges and for decoration, crossing over the top and tied in a bow.

The only known piece of furniture for the doll was a mahogany wardrobe, but three pillow slips suggest that there may have been a bed. Two of the pillow slips, 6 1/16 in X 9 1/16 in (15 X 23 cm), are embroidered with a red M, as were the two handkerchiefs.

One might wonder whether Maggie's favorite colors were red, violet or purple, blue and pink, because they were used repeatedly in the garments she had made for her doll.

The doll Margrete had exquisite accessories of glassware and a toilet set in four boxes (Ill.308, a - d). They display a set of green glassware, three glasses and a matching bottle which still has the label on the bottom: "*Carte blanche mousseux v. Deutz & Geldermann*". A decanter with six matching gold trimmed glasses is similar to the set of English glassware which Maggie's parents had. A beautiful punch set in its original box has six gold rimmed glasses and a lovely bowl fitted into a two tiered circular frame. The fourth box contains a ceramic toilet set of blue banded jars and lids, and a wash bowl and pitcher.

Incidental accessories not usually found for today's dolls (except for Barbies and similar fashion type dolls) includes those considered essential in the doll mother Maggie's family. An attractive little French vanity trunk contains the doll's daily grooming needs: hair brushes and tortoise shell combs, a hair band, and a tooth brush, an ivory powder box, a French labeled perfume bottle and a mirror (Ill.309). Two little pieces of knitting on needles are kept in a knitting bag of red, grey-green, purple, and light yellow stripes. A needle holder contains the other tiny knitting needles. Also of interest are a box of playing cards and a case containing red sealing wax for letters. In a leather coin purse is a 10 ore coin dated 1865. Probably most important to Maggie was the doll's case with small calling cards, 10/16 in X 15/16 in (1.5 X 2.3 cm), inscribed "M.Wolff".

Maggie Wolff had been injured when she visited in England as a young adolescent. Famous doctors were unable to restore her heel, and she was a cripple for the rest of her life. She never married. Maggie died in 1901; the doll was given to her 13 year old niece Ella, who never played with it. Ella married and had three daughters; they did not play with the doll either. When the niece, Mrs. Ella Lichtenberg-Wolff, brought the doll and its lovely wardrobe and accessories to the Danish National Museum in Copenhagen in 1971, the staff was delighted. A wondrous exhibit of many of the items presented here was displayed in a special case on the third floor. And there we met this beautiful English wax doll, which had become completely and aristocratically Danish as "Margrete Wolff".

Ill. 308 a-d

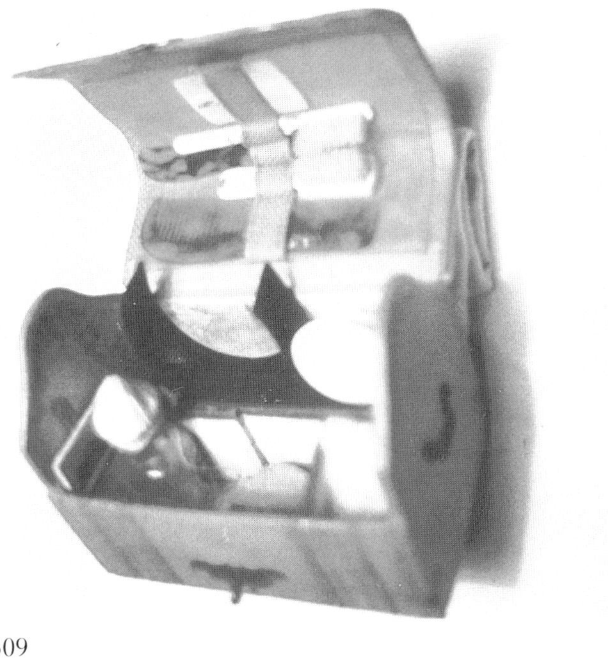

Ill. 309

(starting from top right, counterclockwise)

(308 a - d) Margrete's glassware, and ceramic toiletry:
a. Green bottle and matching glasses.
b. Siphon and decanter.
c. Ceramic toilet set. Photographed at *Nationalmuseet.*
d. Punch set.

(309) Doll's hair grooming trunk. Photographed at *Nationalmuseet.*

Summary

Dolls made in Denmark have been found of many materials: cloth, papier mâché, wood, glazed porcelain and bisque porcelain, metal, composition, celluloid, and plastics. The hunt through literature, museums, private collections, antique shops, and flea markets has been rewarding. Museum shops, and the semiannual toy show, where talented doll artists and knowledgable collectors were gathered, also provided more examples to document, to buy. Geographically the locations of all doll activities dot much of the map of Denmark.

Although doll houses are discussed only incidentally in the museum presentations, doll house dolls represent several materials. They include the wooden Bedste and the pair of the 1920s, the hard plastic trio of Lego dolls, and a variety of bisque dolls reproduced from molds of German originals. Small celluloid dolls from three private collections were shown. Jette Viby's wee cloth dolls for her Millie and the Turn-About doll are of doll house sizes.

The most prolific period of doll production, before today's bonanza of plastic DanDolls and artist dolls, was probably from 1930 to 1960. The celluloid dolls were made during the time of the composition dolls. The earliest dated composition dolls are the 1932 pair in the Randers museum; the earliest celluloids found were the Nisser of the 1930s. Both celluloid and composition doll making peaked in the 1950s. Lack of imports during and after World War II resulted in part in the copying detailed by Finn Risom. However, some lovely Danish designed composition doll heads were thoroughly researched and reported by Mr. Risom.

In contrast with the Danish doll manufacture, play dolls through the years were imported, especially from Germany. Many can be seen in Danish museums. Today's imports include Barbies and many other plastic dolls from HongKong and Taiwan.

A summary of the dolls and toys prominent in the childhood of four generations of the Nielsen family (Ill.310) reveals both Danish and imported dolls. Haldora, the old papier mâché doll of Great Grandmother Nissen, tops this family tree. Her daughter Gunhild shared her well kept composition doll of 1926, a beauty which may be of German origin. The treasured Bodil Marie wears authentic Rømø folk dress more recently made than that of Haldora. Gunhild's daughter Birgitte enjoyed small Steiff animals and trolls and wrote a story about the trolls for the make-believe play with them and her brothers' favored teddy bears. A suitcase filled with teddies of all sizes was taken on an Italian holiday when the youngest child was five years old, and delighted the maids at their hotel. A cherished 6 in (15.2 cm) Ginny and her wardrobe were given to Birgitte in 1961 by Jane Luckey. Because "Ginny understands only English", the new doll mother was obliged to speak to her American doll in that language.

The wives of Birgitte's brothers have kindly shared their childhood dolls for documentation. The yellow haired celluloid doll was given to us by Benedicte. Her report of her favorite, the beautiful silver haired celluloid doll, alerted us to dolls of

ONE FAMILY'S DOLLS AND TOYS

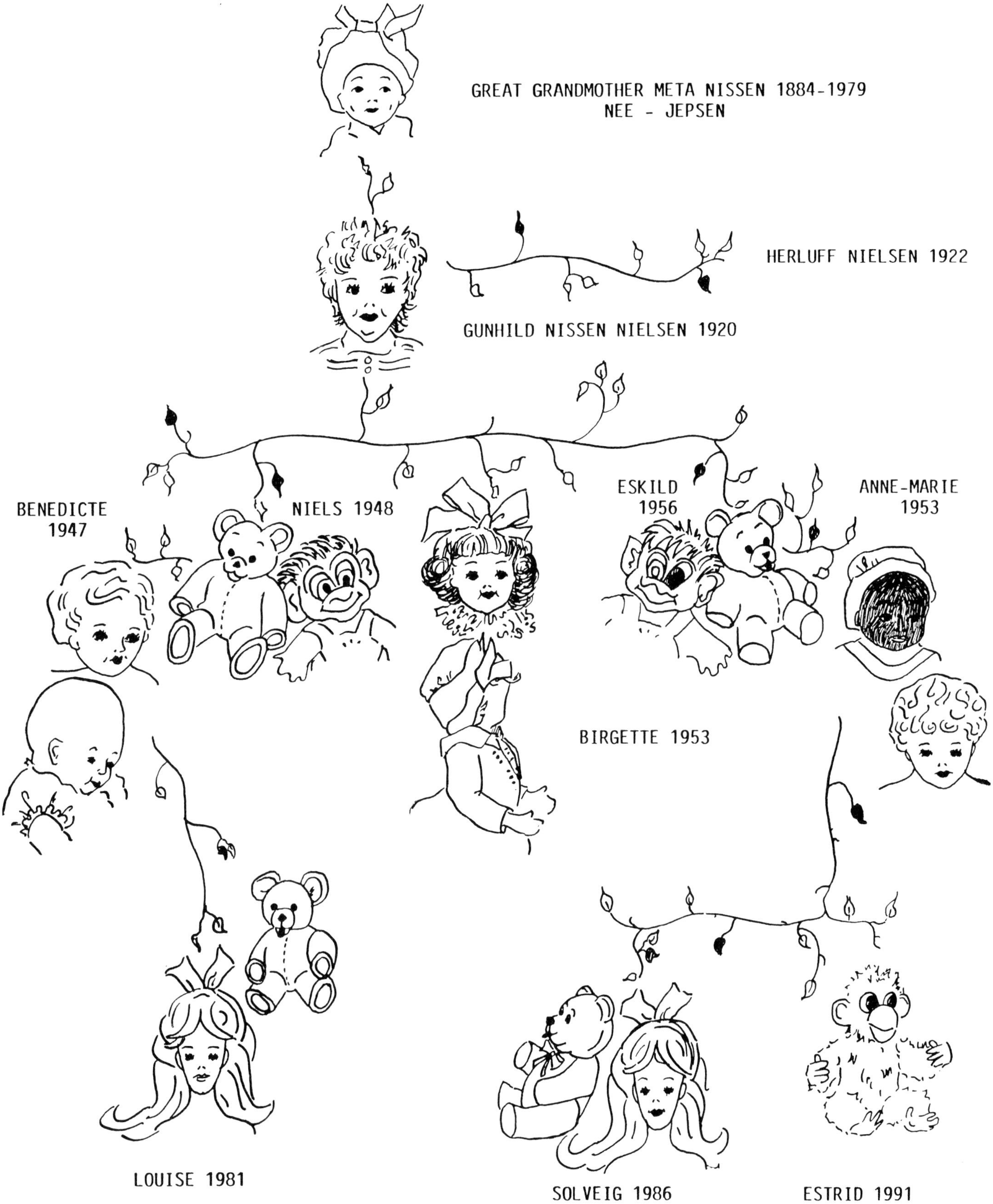

Ill. 310 Family tree of dolls and toys of one family. Sketched by Eileen van Baren

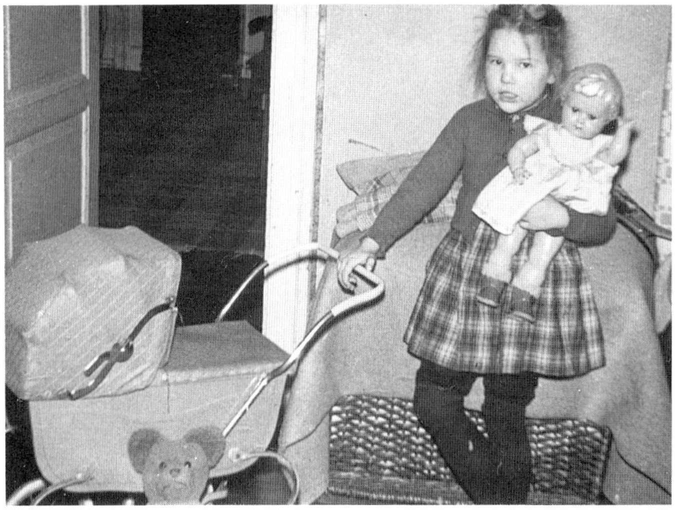

Ill. 311 a

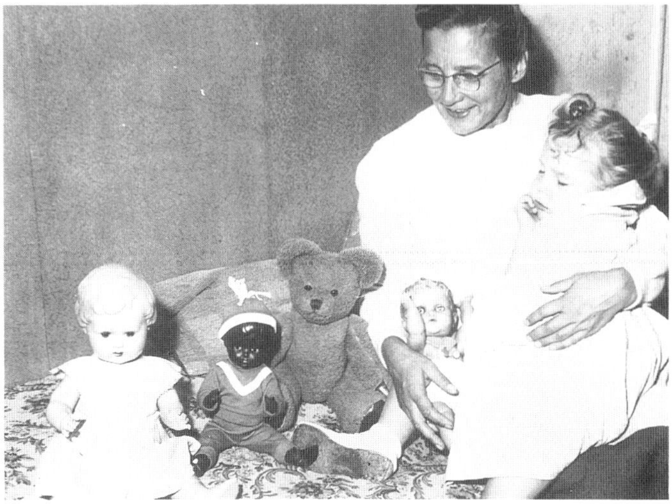

Ill. 311 b

(311 a, b) Dolls and toys of Anne-Marie in 1960s. Courtesy of Anne-Marie Nielsen.

that genre and unique look. The treasured German bisque Dream Baby has a place on the family tree with her Køge celluloid. Anne-Marie's childhood photographs (Ill.311) show her big doll family of celluloids and composition dolls, a black doll probably of rubber, teddies, and a doll carriage. The black doll and her large Køge celluloid doll appear on the tree also.

The fourth generation of three granddaughters are Louise, Solveig, and the baby Astrid. The two older girls are Barbie fans but the old classic teddy bears are equally loved, and the dear Bodil Marie, always a favorite of Aunt Birgitte, is very special to them. From the author's collection two baby dolls, a Horsman and a small Horsman look-alike, were welcomed by the little girls in 1989. The smaller doll was destined to become the Princess on the Pea, in her cigar-box bed filled with mattresses. The baby favors her cuddly monkey.

Dolls and related items found in Denmark constitute an outstanding representative Danish group in the Ora Elizabeth Dolls of the author and photographer. Those include dear old Haldora in her original Rømø costume, and more than a dozen celluloid dolls and heads of the Køge production, from the 4 in (10.2 cm) articulated baby doll to the 23 in (53.4 cm) boy. A collection of Dam trolls includes those of the 1960s and 1970s, a few of the early Dam animals, and animal banks. A small reproduction ByeLo represents the museum of Mme. Siesbye; Lego construction kits containing doll-like figures represent the Lego firm. Small *spraellemaend* found in gift shops and Nisser of several materials add variety to this personal collection.

Wonderful dolls and related items given to us by the artist designers include several cloth dolls by Ester Pedersen, Nisser and *Julemand* candlesticks made by Kirsten Gynther Eriksen, and the exquisite manikin doll from Else Madsen of DanDolls. Jette Vibe's Turn-About dolls were sent by Mr. Erling Andersen, the present manufacturer. And our last surprise, the mascot doll CARLO, was sent from Carlsberg Breweries in December 1990 by Helle Madsen. Most of these Danish dolls will be given to the Danish Immigrant Museum in Elk Horn, Iowa.

This little doll history is primarily a presentation of Danish made dolls which were researched for more than three decades. They are rarely seen in the U.S. The 1840 Royal Copenhagen boy doll head can be seen in the Margaret Woodbury Strong Museum, the only museum in the U.S. of more than twenty queried which has a Royal Copenhagen porcelain doll. Fortunate are the visitors to Copenhagen and Billund museums where more of those exquisite porcelain dolls are displayed. Visits to any of the museums in Denmark are rewarded by the Danes' enthusiasm and hospitality.

Glossary

(Terminology, usage and and spelling of many terms from Danish language, also according to *Webster's Ninth New Collegiate Dictionary*.)

A.M. Initials and mark of dolls by Armand Marseille in Sonneberg, Germany, 1890s to 1930s. Example: Dream Baby.

Armature The support within the doll's body or limbs; can be wire, metal or other material such as hard plastic.

Articulated Having some movement at any of the joints: neck, shoulders, elbows, wrists, hips, knees, ankles.

Articulation The movement of the limbs, usually named in detail of joint(s).

Artist doll One designed and created by the artist, not a copy of another's work.

Artisan An individual who makes dolls with the use of ready made molds or other helps, in contrast with *Artist.*

Artist One who individually creates the designs and parts.

Bisque Unglazed porcelain, with matte finish, contrasted with China. See Glazed porcelain.

Ceramic-bisque From the German *Keramic Bisquit*: A kind of ceramic which is more clay-like than bisque porcelain. Term not used in US; dolls of this substance classified as Composition in American doll book.

Celluloid The early plastic used for dolls from ca 1860. Made of cellulose nitrate and other very flammable materials. Used extensively in Danish doll making, late 1930 - 1960.

Cernit A Danish name for a moldable putty which is oven-baked rather than fired in kiln.

China, or China head The term used in U.S, for Glazed Porcelain doll heads; sometimes limbs were also glazed. Early Danish and German doll heads were called Porcelain.

Composition A mixture of wood pulp or sawdust with binding materials such as plastic, resin, or glue. Used for doll heads, limbs, entire bodies. See Papier Mâché.

Coin box The Danish term for a penny bank.

Coquette Term applied to doll with flirty look and often with ribbon through molded hair.

Doll house doll One of a small size, commonly ca 6 in or less, for play in doll houses or settings. see *Dukkestuedukker.*

Dolly face Term describing happy, smiling girl doll, usually made of bisque or composition.

Dukke, Dukken Danish term for doll, or the doll, often used in combination with other words. See following entry.

Dukkesteudukker Danish term for doll house doll.

Eneret Danish term meaning exclusive rights to use of a design or model.

Ethnic Refers to a doll or figurine representing a specific nationality or race, contrasted with Folk Doll. Most of the series "United Children of our World"(Chapter 9) are ethnic dolls; the Danish and German dolls there would be called "Folk Dolls".

Figurine A statue-like figure with absolutely no articulation. Dolls of the year by Bing and Grondahl were styled after very popular figurines the firm made earlier.

Flirting eyes See Googly.

Folk doll In Denmark, any doll wearing the costume of a specific locality, a city or a region.

Folkedragter Danish term for folk costumes.

Fortune telling doll A doll with fortunes to be selected from folded papers which form the doll's skirt.

Frikadella Hair styled in two braids, coiled over ears in shape of a favorite Danish pastry. See Ram's horns.

Glazed porcelain Hard shiny finish, smooth and glasslike, contrasted with Bisque Porcelain. See China.

Googly Descriptive term for any doll with rougish, side glancing eyes.

Hampleman German word for Jumping Jack. See *Pantin, Spraellamand.*

Historical doll A doll representing a well known political, literary figure, or royalty. Examples: dolls by Kirsten Gynther Eriksen.

Jumping jack Articulated figure of wood or cardboard, activated by pulling string connecting limbs to body.

Julemand Danish name for Santa Claus.

Kaiser baby A nickname given to Kammer and Reinhardt's character Baby, model 100.

K.P.M. The German firm *Koenigliche Porzellan Manufaktur* which made porcelain doll heads in 1840s.

Manikin An articulated figure representing the human body, used to display costumes. Example: "The Fashionables" by DanDoll. (Spelling used in Denmark and according to *Webster's Ninth New Collegiate Dictionary.*)

Mascot doll One made in celebration of a particular sports event. Example: the Dam Troligan.

Merino A breed of sheep with long curly wool used for doll wigs.

Motchmann type Doll bodies copied from a Japanese doll body by German, French. Characterized by stuffed fabric portion of torso allowing movement, especially bowing, a propriety in Japan.

Needle sculpture Artistic creation of doll face etc. in which details of features are outlined with thread and stuffed. Also called Soft sculpture.

Nisse, pl. Nisser Elves of Norwegian and Danish folklore, now the spirited creatures of the Christmas holiday.

Paddle Of wood, hands having a mitten-like shape with no individual finger shaping but may have thumbs. Also called spoon hands.

Pantin French word for Jumping Jack, from French area where the toy originated.

Papier mâché A mixture of paper pulp and some binding material such as glue. Used for doll heads, and doll bodies and limbs. Similar to composition but less durable.

Parian type A white bisque porcelain used for doll heads in 1870-1880s, so-called because of resemblance to white marble of Greek island, Paros.

Porcelain A hard ceramic ware of kaolin clay, feldspar, and quartz. High temperature firing needed. See Bisque, China or Glazed porcelain, Parian.

Ram's horns See *Frikadella* for type of hair styling.

Replica An art object, or a doll, shaped from the original mold. Contrasted with Reproduction. See next entry.

Reproduction doll A doll made from a mold taken from another doll.

S.F.B.J. Initials of *La Societe Francaise de fabrication de bebes et jouets*, a conglomerate of 11 French doll firms, formed on March 6, 1899 in Paris.

Socket neck Construction design in which doll's head fits into hole (socket) between shoulders. Provides articulation.

Soft sculpture See Needle sculpture.

Spadukke Danish name for Fortune telling doll.

Spraellemand Danish word for Jumping Jack.

Tear duct dots Very small red marks at inside corners of doll's eyes.

Three faced doll A doll with three expressions molded in the head; expression changes by turning a knob on top of head. Example: Trudy. See Trio doll.

Topsy turvy doll A doll with head at each end, and skirt which is designed to cover one head while revealing the other doll.

Trio doll Danish name for doll made from mold like (or of) American Trudy doll. See Three faced doll.

Troll Mythical creature of Scandinavian legend. Designed by Thomas Dam, manufactured from 1959 through today.

References

Introduction

[1] Theriault, Florence, Oct.22, 1989, *Dolls: The Early Years 1780 - 1880*, Theriault's, the Dollmasters, Annapolis, Maryland, Forward.
[2] Tvermoes, Susanne, 1978, *Gamle dukker*, SESAM A/S,Aarhus.
[3] Sigsgaard, Jens and Varnild, Og, 1982, *Det Legede Vi Med*, Nyt Nordisk Forlag Arnold Busck, Copenhagen.
[4] Merrill, Madeline Osborne, 1985, *The Art of Dolls*, Robert Hale, London. Reissued 1985, Hobby House Press, Inc., Cumberland, MD.
[5] Lorenzen, Erna, 1967, "En dukke fra 1830 - 40", in *Kobstadsmuseet "Den Gamle By"*, Arborg, (the yearly publication), Aarhus, 101 - 110.

Chapter 1

[6] Coleman, Dorothy S., Elizabeth A., and Evelyn J., 1968, *The Collector's Encyclopedia of Dolls, Vol.I*, Crown, 534 - 535.
[7] Coleman, Dorothy S., Elizabeth A., and Evelyn J., 1986, *The Collector's Encyclopedia of Dolls, Vol.II*, Crown, 1014 - 1015.
[8] Dröscher, Elke, 1982, *Puppen im Wandel der Zeit*, Keyser, Munich, 46 - 47.
[9] Tvermoes, 1978, op cit., 18.
[10] 1977, *The Book About Little People In Legoland's Doll Collection*, LEGOLAND A/S.
[11] 1987, *THE LEGOLAND DOLL COLLECTION*, LEGOLAND A/S.
[12] Merrill, op cit., 82, 83, 88, 94, 110.
[13] Dröscher, op cit., 46 - 47.
[14 a-e] Merrill, op cit., a, b: 82, 88, c. 94, d. 110, e. 83, 90.
[15] Fawcett, Clara Holland, 1949, *On Making, Mending and Dressing Dolls*, H. J. Lindquist Pub., reprinted Hobby House Press, Inc., Cumberland, MD., 145.
[16] Poulsen, Van, 1976, *Danish Painting and Sculpture , Det Danske Selskab*, Copenhagen, 49.
[17] Winther, Christian, 1830, reissued 1985, *Flugen til Amerika* (*The Flight to America*), Glydendal, Denmark,1.
[18] Dröscher, op cit., 45, 59.
[19] 1989 *CATALOGUE*, ROYAL COPENHAGEN, 20.

Chapter 2

[20] Knudsen, Inge Holm, 2-21-1988, *"I fornemt selskab"*, ("A distinguished society", or "....collection"), *Morgenposten* (*Morning Post*), 45.
[21] Ibid.
[22] Selbie, Robert, 1977, *The Anatomy of Costume*, Crescent Books, New York, 40.
[23] Lauring, Palle, 1960, *A History of The Kingdom of Denmark*,Host & Son, Copenhagen, 145.
[24] Selbie, op. cit., 43, 46 - 49.
[25] Lauring, op. cit., 152.
[26] Selbie, op. cit., 46, 49, 51.
[27] Kroyer, Susanne, 3-25-1989, *"Hun lever og ander for sine dukker"* ("She lives and breathes for her dolls"), *Nyborg Avis* (*Nyborg News*), 17.

Chapter 3

[28] Fraser, Antonia, 1984, *Schöne Puppen* (*Beautiful Dolls*) Battenberg Verlag, Munich, 114.
[29] Kragelund, 1978/1989, *Folkedragter*, (Folk Costumes), Lademann, Copenhagen, 84 - 91.
[30] Fanø Turistbureau, 1989 *Welcome Fanø*.
[31] Hathaway, Mary, Sept./Oct. 1980, "Around the World with Dolls (Denmark)", *National Doll World*, 12 - 13.
[32] Kragelund, op.cit., 50 - 51.
[33] National Travel Association of Denmark, 1952, *Denmark Cavalcade 1952*, Forlaget Aktuelle Boger, 6.
[34] National Travel Association of Denmark, 1965, *Denmark Cavalcade 1965*, Forlaget Aktuelle Boger, 31.
[35] Undated brochure, "Royal Copenhagen Porcelain Ceramic Art".
[36] Nedergaard, Pamela, Descriptive enclosure with doll, personal communication.
[37] Kragelund, op.cit., 41.
[38] Kragelund, Ibid.

Chapter 4

[39] May 1989, personal communication, Niels Dam.
[40] 1988, *Portrait of a Successful Family*, Company prospectus, 1. Reprinted with permission of Mr. Niels Dam.
[41] Op cit, 7.
[42] Smith, Patricia, 1976, *Modern Collector's Dolls, Third Series*, Collector Books, Paducah, KY, 267.
[43] Smith, Patricia, 1979, *Modern Collector's Dolls, Fourth Series*, Collector Books, 240 - 241.
[44] Smith, Patricia, 1984, *Modern Collector's Dolls, Fifth Series*, Collector Books, 264 - 267.
[45] Gibbs, Patikki and Tyson, 1987, *The Collector's Encyclopedia of Black Dolls*, Collector Books, 1987, 27.
[46] Schroeder, Bill, 1976, *The Standard Modern Doll Identification & Value Guide*, Collector Books, Paducah, 66.
[47] Smith, Patricia, 1980, *Doll Values Antique to Modern, Series II*, Collector Books, Paducah, 274 - 275.
[48] Wilson, Jacqueline, Aug./Sept. 1988, "Trolls areTerrific", *Doll Reader*, 233 - 235.
[49] Wilson, Jacqueline, Feb.1990, "Tracking the Troll", *National Doll World*, 38 - 41.
[50] Umlah, Terry Lynn, Feb./March 1990, *Women's Circle Doll Designs*, 40 - 43.
[51] Wilson, 1990, op. cit., 43.

Chapter 5

[52] 1950, *Historical Atlas*, C.S. Hammond & Co., New York, NY, H-20, H-22, H-31.
[53] Sigsgaard and Varnild, op cit., 79.
[54] Brandt, Christina, 1989, *Nisserne pa Grønland*, Pobra a/s, Gentofte, Denmark.

Chapter 6

[55] Sigsgaard and Varnild, op cit., 61.
[56] Nellemann and Thompsen, May 1990, *International Legetojsauktion 596*, 33.
[57] Sigsgaard and Varnild, op cit., 62.
[58] Groninger, Margaret, Winter 1983, "Wee Immigrants, Part II", *Doll News*, 47 - 48.
[59] Sigsgaard and Varnild, op cit., 62.
[60] *Villabyerne* (a Gentofte newspaper), Jan.18, 1989 "*Dukker i fine dragter som sy-selv-opgave*", 4. ("Dolls in fine dress become her mode of self-expression".)

Chapter 7

[61] Nielsen, G., personal communication.
[62] Tarnowska, Maree, 1987, *Rare Character Dolls*, Hobby House Press, Cumberland Park, MD, 16; 82.
[63] Risom, Finn, 1990, personal communication.

Chapter 8

[64] Winter 1982, Luckey, Pauline, , "German Dolls, Then and Now, Part III: Schildkröt", *Doll News*, 70.
[65] Undated, Andersen, Kirsten and Andersen, Harry, "A History of a Doll Factory", from a Danish manuscript, translated by Pamela Nedergaard.
[66] 1966, *Encyclopedia Americana*, Vol.8, 1679.
[67] 1990, Maxine Bach, Museumsinspektor, Køge Museum, citing *Dagbladet*, 14/6/1963, Personal communication.

Chapter 11

[68] 1982, *The Legoland Doll Collection*.
[69] Luckey, Pauline, translated by Gauder-Bonnet, Ursula, Jan. 1985, *Puppen und Spielzeug*, 46 -47.
[70] Sigsgaard and Varnild, op cit, 61.
[71] Merrill, op cit, 415.
[72] Bache, Camme, Brochure.
[73] Bache, 1990, personal communication.
[74] Luckey, Pauline, Summer 1990, *Doll News*, "Blue Eyed Black Dolls . . .A Phenomenon", 62.
[75] Sigsgaard and Varnild, op cit, 85 - 86
[76] Ibid, 310 - 311.
[77] 1950, Advertisement, Tekno, courtesy of I. and P. Pedersen.
[78] Reinelt, Sabine, 1986, *Puppen und Spielzeug aus Zelluloid*, (*Dolls and Toys of Celluloid*), Weingarten, Freiburg, Germany, 108 - 109.
[79] Sigsgaard and Varnild, op cit, 288.
[80] Ibid, 290.
[81] Ibid, 290
[82] Ibid, 289.
[83] Madsen, Helle, 1990, personal communications.
[84] Information on doll's carton
[85] Ibid.

Chapter 12

[86] June 1990, Nedergaard, Pamela, personal communication.
[87] Oct. 1989, Brevik, Aase, personal communication.
[88] 1990, Bache, personal communication.

Chapter 13

[89] von Pein, Lis and Lassen, Hans, 1977, *The Book About Little People in Legoland ®S COLLECTION*, Legoland A/S, 3.
[90] Jan. 1978, "Titiana's Palace", in *Puppen und Spielzeug*, 18 - 20.
[91] Ganzhorn, Hanne, and Damm, O. Thygesen, 1987, *Titiana's Palace*, Legoland A/S, 2.
[92] Wall poster at Legoland.
[93] Poulsen, Doris, Oct.6, 1989, personal communication.
[94] Kjaer, Birgitte, Sept.27, 1989, personal communication.
[95] Oct.1989, *Copenhagen This Week*, Forlaget Folia ApS, 41.

Chapter 14

[96] Kragelund, Minna, 1974, *Dukken Margrete, The Doll Margrete, Nationalmuseet*, 5.
[97] King, Constance Eileen, 1977 and 1978, *The Collector's History of Dolls*, Robert Hale Ltd, London, and St. Martin's Press, New York, 274, 278.
[98] Op cit., 276.
[99] Gauder, Gert, 3/1978, "*Zur Geschicte Der Wachspuppen*" in *Puppen und Spielzeug*, 4, 10, 11.
[100] Kragelund, Loc cit., 10.
[101] Op cit, 47.